RACE AND ENVIRONMENTAL JUSTICE IN THE ERA OF CLIMATE CHANGE AND COVID-19

Environment, Health, and Well-being

This series tackles the relationship between health and the environment, paying particular attention to changes taking place over time and across place. It seeks to illuminate the causes and consequences of human, more-than-human, and environmental ill-health, while also attending to possibilities for well-being, flourishing, and repair. Encouraging an expanded notion of health, Environment, Health, and Well-being presents scholarship that considers human well-being as directly correlated with health systems, extends the notion of health and well-being beyond the purely human frame, and interrogates planetary health through specific landscapes, ecologies, and human and more-than-human activities. Recognizing the ecological, political, social, and viral turbulence of our current times, the monographs and edited collections in this series look to interdisciplinary practice within the field of the environmental humanities as a way of understanding the present, reflecting upon the past, and rethinking possibilities of the future.

Environment, Health, and Well-being, while grounded in the environmental humanities, understands the barriers to environmental health as tied to legacies of extraction, consumption, colonization, and unlimited growth. It is thus especially interested in scholarship from indigenous studies, race studies, gender and queer studies, and disability studies, as well as approaches that address histories and futures of labor and profit. Environment, Health, and Well-being welcomes projects from new and established scholars, in and outside academia, which make visible for audiences the timeliness and necessity of interdisciplinary research on the relationships between humanity and environments. The contributions in this series capture the multifaceted nature of environmental health and foreground the importance of perpending the planet's well-being in these ecologically precarious times.

RACE AND ENVIRONMENTAL JUSTICE IN THE ERA OF CLIMATE CHANGE AND COVID-19

Edited by Tatiana Konrad

MICHIGAN STATE UNIVERSITY PRESS | *East Lansing*

Michigan State University Press
East Lansing, Michigan 48823-5245

Published with the support of the Austrian Science Fund (FWF): 10.55776/PUB1151

FWF Austrian Science Fund

Research results from: Austrian Science Fund (FWF): 10.55776/P34790

Library of Congress Cataloging-in-Publication Data
Names: Konrad, Tatiana, 1991– editor.
Title: Race and environmental justice in the era of
climate change and Covid-19 / Edited by Tatiana Konrad.
Description: East Lansing : Michigan State University Press, [2025] |
Series: Environment, health, and well-being | Includes bibliographical references.
Identifiers: LCCN 2024057183 | ISBN 9781611865288 (paperback)
| ISBN 9781609177799 | ISBN 9781628955439
Subjects: LCSH: Environmental justice. | Racial justice. | Climatic changes—Social aspects.
| Human ecology. | COVID-19 Pandemic, 2020–2023.
Classification: LCC GE220 .R33 2025 | DDC 363.7/05614—dc23/eng20250115
LC record available at https://lccn.loc.gov/2024057183

Book design by Anastasia Wraight
Cover design by Amanda Frost
Cover art: Photo depicting a bright green lichen on the old stone wall by alicefoxartbox, Adobe Stock.

Visit Michigan State University Press at *www.msupress.org*

Contents

vii   Naming the Crisis: Conceptualizing Environmental, Climate, and Health Injustices through Race and Indigeneity, *Tatiana Konrad*

PART 1. **Air, Breath, and (In)Visibility of Environmental Racism**

3   Atmoracism: Air and Precarities of Health, Environment, and Race, *Tatiana Konrad*

23   The Respiratory Politics of Air-Breath: India's Epidemic Intensities, *Bishnupriya Ghosh*

47   Fueling Toxicity: Fast Fashion, Air Pollution, and Slow Violence, *Savannah Schaufler*

PART 2. **Health, Sustainability, and Race**

73   Environment and Health: The Impact of Historical Environment on Inuit Qanuinngitsiarutiksait in the Era of Anthropogenic Climate Change, *Jeevan Stephanie Kaur Toor, Tagaaq Evaluardjuk-Palmer, and Josée G. Lavoie*

107   COVID-19 and Socioenvironmental Sustainability: Grassroots Strategies of Autonomy and Healing among and between the Ngigua in San Marcos

Tlacoyalco, Puebla-Mexico, *María Cristina Manzano-Munguía, Guillermo López Varela, and María Sol Tiverovsky Scheines*

131   The Stop Cop City Movement: Environmental Racism, Community Policing, and Necropolitics in Post–COVID-19 Atlanta, *Juan M. Floyd-Thomas*

PART 3. Decolonizing and Implementing Environmental Justice

169   Material Flows in Landscapes of Injustice, *Nikiwe Solomon*

183   Future Tense: The Role of Race, Risk, and Environmental Justice, *Helen Bond*

207   Decolonizing Environmental Justice: Centering Black and Indigenous Solutions to the Climate Crisis, *Autumn Asher BlackDeer and Sierra Roach Coye*

231   Conclusion: Decolonizing Environmental Justice Pedagogy, *Tatiana Konrad*

235   About the Contributors

241   Index

Naming the Crisis

Conceptualizing Environmental, Climate, and Health Injustices through Race and Indigeneity

Tatiana Konrad

Social justice is tightly linked to many other forms of justice. When studying the effects of climate change, COVID-19, and environmental transformation on Black, Indigenous, and People of Color (BIPOC) communities around the world, it is crucial to understand that these types of injustice selectively target people of color because of the lack of social justice, compounding the inequality and oppression that the people are already experiencing. Social injustice both provokes and sustains other forms of injustice, of which environmental, climate, and health injustices are prime examples.

Scholars from various disciplines have energetically engaged with the issues of social and environmental justice.[1] Informed by this important research, *Race and Environmental Justice in the Era of Climate Change and COVID-19* is both a scholarly contribution to the ongoing discourse and a form of activism for environmental, climate, and health justice. Using race and indigeneity as an analytical lens, the book explores how justice in the era of climate change and COVID-19 is envisioned, depicted, and achieved. Examining the relationships among humans and environments, this edited collection, in line with multispecies justice scholars, recognizes that it is crucial to broaden "the rules of inclusion that constitute politics" and calls for "new political imaginaries that take into account the ontological diversity, relational complexity, and incommensurable forms of communication and desire, within which just arrangements and outcomes can be cocrafted."[2] While this book largely focuses on humans and environments, its explorations of (in)justice help not only to illustrate the wide health and safety gap between individuals, communities, and even nations living under different environmental conditions, but also to move beyond the human toward justice for all. Instruments of "culture, politics, knowledge, and communication," which multispecies justice scholars see as key in securing justice for the multiple species that inhabit and constitute the Earth, are also central to fighting inequality and injustice among humans and for envisioning and securing sustainable futures for all.[3]

This book operates along lines of social, racial, environmental, and health justice, exploring several research trajectories. It opens with "Part 1: Air, Breath, and (In)Visibility of Environmental Racism." Here, the contributors review environmental injustice through the lens of environmental racism. Specifically drawing on air, breath, respiratory problems, and air pollution, this section demonstrates how environmental injustice can manifest in (in)visible ways, targeting BIPOC communities in both the Global North and Global South.

This section opens with Tatiana Konrad's "Atmoracism: Air and Precarities of Health, Environment, and Race." Konrad coins the concept of *atmoracism* to describe oppression that BIPOC communities experience through air. Atmoracism—some examples of which include exposure to air pollution, the airborne COVID-19 virus, and police brutality that interferes with the breath of Black bodies—according to Konrad, is a form of methodical, continuous violence against people of color. This essay, therefore, sheds light on yet another form of structural oppression that, due to its connection to the allegedly "invisible" air, might remain unnoticed and neglected by scholars and other activists. Seeing air and recognizing its materiality can help identify cases of atmoracism and, through this, fight for racial equality and social justice. This type of social justice is inseparable from environmental justice, health justice, and racial justice. This essay thus examines the precarious conditions in which BIPOC communities exist, theorizing atmoracism through the lens of environmental–health–racial justice. It draws on PBS's *Hindsight* (2021) to demonstrate how BIPOC communities use air space to fight oppression, a phenomenon that Konrad terms here as *atmoactivism*.

In "The Respiratory Politics of Air-Breath: India's Epidemic Intensities," Bishnupriya Ghosh argues for respiratory justice due to the trouble with air-breath during the COVID-19 pandemic. Analyzing images of air-breath that emerged in the human-caused disaster that was India's second wave (April–May 2021), this essay explores the potential of a political aesthetic that leans toward environmental justice. India's second wave exemplifies the stark differences of experience during the otherwise globally synchronous pandemic. The logic of molecular colonialism surfaced yet again during COVID-19, as those with long-term lack of access to health care, nutrition, and housing died in greater numbers all over the world. Within this larger global story, Ghosh plots the catastrophic effects of the Modi-Shah government's administrative failures on India's most vulnerable communities: during the second wave's medical oxygen supply crisis, for instance, those with means could commandeer manufactured air while others perished without, their passing visible in the smoke from crematoria. How might the afterimages of situated pandemic experiences compel a robust respiratory politics? Ghosh argues for "epidemic intensities" as an analytic that opens up differentials in the COVID-19 air-breath experience and offers a critical method for tracking such intensities from the standpoint of visual culture studies. The outcome is a situated political aesthetic for Delhi's COVID-19 experience.

The section concludes with "Fueling Toxicity: Fast Fashion, Air Pollution, and Slow Violence," where Savannah Schaufler maintains that with ninety-two million tons of garments ending up in landfills worldwide, there is no doubt that the constant production and consumption of fashion has negative social and environmental consequences. Regardless of the location of their original entry to the global market, the majority of textiles end up in one of the world's largest secondhand markets, the Kantamanto Market, located in Accra, Ghana. Here, masses of clothes from clothing bins and charity collections mainly from the United States and Europe, especially Austria and Germany, are sent for resale and reuse. In other words, while Western countries attempt to recycle their fast-fashion textiles, Ghana bears the burden of textile pollution. There, mountains of worthless and mainly inferior-quality clothing, consisting of plastic fibers that cannot be reused or resold, are incinerated, polluting the air with noxious fumes, especially from the synthetic chemicals and dye toxins. Referencing concepts such as actor-network theory, Zsuzsa Gille's "waste regimes," and Rob Nixon's "slow violence," this essay explores toxic embodiments, examining the intricate interplay of time, space, and the body within the fast-fashion industry, while considering its colonial legacies, power dynamics, and their impact on aerial health across populations. Taking COVID-19 as a flashpoint for rethinking relationships to air—as this pandemic made aware the physical presence of air—this essay presents the multiple and complex dependencies between production and disposal of products. This allows for a radical reflection on the globalized waste of clothing, with particular attention to air pollution, health impacts, and questions of agency.

In "Part 2: Health, Sustainability, and Race," the contributors provide a deeper investigation of the issue of health as it intersects with the environment and race. The global focus of this section helps adumbrate similarities in how race and health have become intertwined in the era of environmental crisis in different parts of the world.

In "Environment and Health: The Impact of Historical Environment on Inuit Qanuinngitsiarutiksait in the Era of Anthropogenic Climate Change," Jeevan Stephanie Kaur Toor, Tagaaq Evaluardjuk-Palmer, and Josée G. Lavoie explore the impact of historical environment (i.e., colonialism) on qanuinngitsiarutiksait (good health and well-being) in Inuit communities in northern Canada. The erosion of Inuit social and cultural fabric has made it harder to adapt to climatic changes, the harms of which have been exacerbated by existing socioeconomic inequalities. Climate change affects qanuinngitsiarutiksait, causing poorer mental health, food and water insecurity, increased disease, ice-related accidents, and increased impact of persistent organic pollutants. This essay aims to move from deficit to strengths-based research by exploring Inuit adaptations to anthropogenic climate change, centering Inuit voices and initiatives. The essay discusses the Canadian government's approach to climate change in Inuit communities, with recommendations for policy and research from an Inuit perspective. Toor, Evaluardjuk-Palmer, and Lavoie argue that the maintenance of qanuinngitsiarutiksait requires a holistic method where Inuit Qaujimajatuqangit (Inuit knowledge) is considered and researchers collaborate with Inuit people; further work is also needed to address the ramifications of the colonial past. Their aim is to commit to reversing the effects of anthropogenic climate change in the Arctic and to supporting those impacted by colonial actions. This research provides a novel contribution through its basis in Inuit knowledge and concepts, "historical environment," and the synergy of the theoretical approaches of anthropology and geography.

Following this essay is "COVID-19 and Socioenvironmental Sustainability: Grassroots Strategies of Autonomy and Healing among and between the Ngigua in San Marcos Tlacoyalco, Puebla-Mexico." Here, María Cristina Manzano-Mun-guía, Guillermo López Varela, and María Sol Tiverovsky Scheines claim that the socioeconomic and health crises caused by the COVID-19 pandemic in Mexico have especially affected its most vulnerable populations, including, but not limited to, Indigenous communities, women, children, and youth from poor rural areas. Over the past three years, COVID-19 intensified the need for improving the health-care systems, medical services, and personnel (e.g., intensive care units and medications), but it also demonstrated the need for food supplies, employment, and social security across the country. Specifically, the pandemic prompted Indigenous communities to seek autonomy in food production and alternative health-care systems. This essay looks at how the Ngigua community implemented grassroots strategies of autonomy through socioenvironmental sustainability projects that included, but were not limited to, the use of traditional medicine, the rain petition, and the "sowing" of rain, while caring for the community *jagüey*—a natural water reservoir used for daily communal needs—vis-à-vis COVID-19 restrictions and mortality effects.

This section concludes with Juan M. Floyd-Thomas's "The Stop Cop City Movement: Environmental Racism, Community Policing, and Necropolitics in Post–COVID-19 Atlanta." In the wake of 2020, the Atlanta Police Foundation and the municipal government proposed a $90 million police training facility derisively called "Cop City." Initial plans for Cop City were sparked by a startling uptick in major crimes coinciding with the onset of the COVID-19 pandemic, not only in the Atlanta metropolitan area but also in countless other cities across the United States. The proposed training center was designed to be a mock urban space replete with homes, convenience stores, and even a nightclub, meant to occupy a heavily forested area in Atlanta's southwestern section. However, following the January 2023 murder of queer, Indigenous-Venezuelan activist Manuel "Tortuguita" Páez Terán by local police officers, an ad hoc, eclectic coalition of residents, police reform activists, environmentalists, and advocates for the homeless engaged in a prolonged confrontation with the police at the construction zone. The police responded to the outburst during an activist-organized "Week of Action" by rushing to the planned demonstration while many members were attending a music festival in Páez Terán's honor. This feud peaked when hundreds of activists breached the construction site, burned police cars and construction vehicles, and damaged the property's infrastructure. This resulted in the arrest of some fifty protesters, many of them being accused of domestic terrorism. How did a woodland area on the outskirts of Atlanta, Georgia, become an unlikely new battleground in debates over residential hypersegregation, environmental racism, and community policing reform? To better fathom this current predicament, Floyd-Thomas argues, we need to grasp what philosopher Achille Mbembe calls "necropolitics," described as "the ultimate expression of sovereignty . . . the power to dictate who may live and who must die."[4] Drawing on social science research, policy analysis, and archival materials, this essay reveals COVID-19's complex impact on Atlanta's public health and social welfare, ranging from urban deforestation to police brutality.

Finally, "Part 3: Decolonizing and Implementing Environmental Justice" provides practical solutions to environmental inequality. Drawing on the voices of representatives of BIPOC communities in both the Global North and Global South, the essays in this section explore cultural, social, political, and other changes that must take place to ensure environmental justice.

Nikiwe Solomon's exploration of environmental injustice in South Africa in "Material Flows in Landscapes of Injustice" opens this section. A common theme in the early months of the pandemic was that COVID-19 has made visible the fault lines of inequality in society. However, for many across the globe, these fault lines were always visible in the toxic environments where they reside. BIPOC and poor communities have often been forced to live in historically segregated areas, where highly toxic materials flow through water, soil, and air. In the age of the Anthropocene, a new epoch in which we recognize the changes in how pathogens, molecules, gigabytes, and currencies flow through human intervention, the question that comes to the fore is what modern governance would look like if flows are central to how we imagine the future in a time of growing climate change. Currently, climate interventions, legal structures, and social and environmental governance approaches to the climate crisis are not adequately equipped to deal with the global flows that transcend particular ideas of territory, space, and time. What is needed, Solomon suggests, is a form of governance that recognizes the interrelationships between the geologic, the biotic, and the social. Solomon proposes a shift in governance from one that focuses on control, command, and prediction to one that focuses on human–nature entanglements through material flows, a form of governance that is connected to what is happening on the ground rather than working from outside the socioecological worlds it is meant to serve. This approach could inform policy that seeks to reconnect rather than increasingly separate the human from nature.

The section continues with Helen Bond's "Future Tense: The Role of Race, Risk, and Environmental Justice." This essay addresses the intersections between environmental injustice and perceptions of race, risk, and visions for the future. Responding to risk is dependent on how we perceive and understand that risk. Risk communication during crises like the COVID-19 pandemic and climate change will require trust and behavior change. How we perceive and understand risk can partly impact our willingness to change our behavior in response to it. In this essay, Bond reports on a study that measures students' knowledge and understanding of environmental issues and associated risks, and their willingness to engage in pro-environmental behavior (PEB). PEB is any behavior or action that helps or benefits the environment or does little harm.[5] For the purposes of this study, taking action is defined as identifying opportunities for engagement and initiating informed and ethical action.[6] Reasons for environmentally friendly behaviors can differ widely across social groups and situations. A number of factors are thought to predict agency around PEB, including risk perception and social cohesion.[7] People are generally concerned about environmental problems because of the negative consequences that may result from them. The perception of injustice and one's relationship to the environment are additional drivers of concern and action.[8] One's relationship to the environment is another important driver of environmental concern. The Environmental Concern Scale was used to measure the degree of importance students place on certain environmental problems due to their perceived risks. It is worth noting that not all consequences merit equal concern. People differ in what they find worrying, and ultimately what motivates them to change their behavior. How connected we feel to nature and our community can be predictive of engaging in PEB. This essay shares research findings regarding race, risk perception, environmental concern, and the concept of future tense—a mitigation strategy for Black youth envisioning futures in an unequal world. Racism and discrimination in the lives of Black youth undermine their ability to be future positive. Black adolescents are exposed to racial discrimination in the United States at levels that can thwart their ability to imagine a sustainable and just future.

This section concludes with Autumn Asher BlackDeer and Sierra Roach Coye's "Decolonizing Environmental Justice: Centering Black and Indigenous Solutions to the Climate Crisis." Environmental injustices such as the climate crisis are inextricably tied to the ongoing processes of colonialism, dispossession, and racial capitalism. This essay critiques so-called solutions to the white climate crisis that are grounded in colonial, anthropocentric thinking and perpetuate systemic harm against BIPOC communities. The authors argue that these Western conceptions of environmental justice fail to encapsulate what BIPOC communities consider justice. BlackDeer and Coye center Indigenous cosmologies and abolition ecologies in an attempt to decolonize the environmental justice movement. Grounded in their lived experiences as queer Black and Native scholars in conversation with scholarly literature, this work delineates historic impacts of colonialism on BIPOC communities, demonstrates the continued coloniality within the present-day environmental justice movement, and introduces Indigenous cosmologies and abolition ecologies as means to decolonize it. Ultimately, BlackDeer and Coye claim that there can be no environmental justice without Black liberation and Indigenous sovereignty.

Finally, in "Conclusion: Decolonizing Environmental Justice Pedagogy," Tatiana Konrad gestures to other sites of inquiry, the outcomes of decolonial environmental justice, and what readers may want to consider next. Konrad also claims that it is crucial to include environmental justice in curricula and emphasizes that *Race and Environmental Justice in the Era of Climate Change and COVID-19* can be used to teach race and environmental justice.

Recently, scholars working in environmental justice, and multispecies justice in particular, have called for "a *transformation* in justice," emphasizing the need to reconsider the ways both "the subject of justice" and "the telos of justice itself" are understood.[9] Part of this work is to recognize race as the key lens through which injustice can be traced, identify the exact locations and people—individuals, communities, and nations—affected by environmental conflict, and outline clear ways in which environmental justice can be restored, including via the eradication of environmental racism. Regions and nations in the Global South experience the effects of climate injustice with full force, while at the same time they largely develop and sustain the Global North through natural resource extraction.[10] This process of unequal, corrupt, and sometimes even criminal extractivism for the economic benefit of the Global North degrades Earth's major domains, including air, water, and soil, triggering environmental crisis. Literary scholar Jennifer Wenzel claims that the consequences of such extractivism are "a borrowing against—even theft of—other people's futures."[11] It is thus essential to not only understand the real ramifications of environmental racism and inequality on affected individuals but also include those individuals in the processes of decision-making, particularly in cases when justice can be, or already is, sabotaged. Such collaborations can open working pathways to address grave issues.[12] For example, there are Indigenous scholars and activists who oppose "the future-oriented temporality of justice in Western paradigms," emphasizing that the current neglect of violence and injustice will not bring communities to any just future.[13] Taking into account and relying on these views as we work toward environmental solutions can, indeed, lead to justice for all.

It is crucial to consider class, gender, sexuality, ethnicity, race, and (dis)ability in environmental issues.[14] Diversity studies scholars like Karen Bell emphasize that, for example, "if the environmental movement can be more inclusive, it can improve its influence on policy and society."[15] Who are the actors (both those who intensify or maintain environmental degradation and those whose lives, health, and well-being are directly affected by it) and what events exactly (instances of environmental injustice, environmental conflicts, etc.) become popularized? This is an important question through which it is possible to identify very tangible acts of environmental injustice. At the same time, it is significant to understand that environmental injustice is not a unified phenomenon, but rather it represents a variety of mistreatment, conflict, and the effects thereof, and thus must be addressed appropriately to each individual case. Similarly, scholars from the Global South describe environmentalism as different in every country, with "its own unique characteristics and trajectories."[16] Environmental injustice is a global issue, and only collective action can solve it.[17] Scholars insist on "the urgent need for North–South collaboration" that can help solve or at least minimize the environmental crisis.[18] Yet environmental justice cannot be restored through one method; it is a task of multiple activities that involve different people, inspect different cases, and, ultimately, fix different problems.

Race and Environmental Justice in the Era of Climate Change and COVID-19 draws on the issues of both justice and injustice: "Injustice (the lack of something) is often more tangible than justice (the supposed fullness or perfection of something)."[19] To understand justice, one needs to fully comprehend the multiple forms of injustice experienced by select individuals and groups; and, vice versa, to be able to address injustice, it is crucial to recognize the true meaning of justice. For this purpose, scholars emphasize the need to "nam[e] forms of justice and injustice," including, for example, "social justice, restorative justice, and distributive justice," in order to be able to "produce a sense of action and agency."[20]

The world is in crisis. Yet this crisis is experienced by individuals very differently, depending not only on one's location but also on one's skin color. Environmental and health issues have become intricately connected to the issues of social justice. It is crucial to decolonize languages, behaviors, attitudes, and very real political actions that discriminate against and harm BIPOC communities. Contributing to the ongoing debates, this book supports the claim that "functioning environments are a necessary condition for the fulfilment of other, intersectional justices—like environmental justice, social justice, and racial justice."[21] Environmental justice scholars Thom Davis and Alice Mah note: "It is difficult to make sense of a historical moment when you are caught in the middle of it—and difficult to tell if it even is a moment, or just a small part of something far bigger."[22] Whether the historical change is already happening or humanity is on its way to making this change, *Race and Environmental Justice in the Era of Climate Change and COVID-19* is a scholarly endeavor to foreground the voices from world communities, provide solutions to environmental and health crises, and restore justice for all.

NOTES

1. The following works, along with further sources quoted in this introduction, provide a solid overview of the ongoing research on environmental justice: Adriana Allen, Liza Griffin, and Cassidy Johnson, *Environmental Justice and Urban Resilience in the Global South* (New York: Palgrave, 2017); Robert D. Bullard, introduction to *Confronting Environmental Racism: Voices from the Grassroots*, ed. Robert D. Bullard (Boston: South End Press, 1993), 7–14; Munamato Chemhuru, *Environmental Justice in African Philosophy* (London: Routledge, 2022); Veronica Maria Sol Herrera, *Slow Harms and Citizen Action: Environmental Degradation and Policy Change in Latin American Cities* (Oxford: Oxford University Press, 2023); Ryan Holifield, Jayajit Chakraborty, and Gordon Walker, "Introduction: The Worlds of Environmental Justice," in *The Routledge Handbook of Environmental Justice*, ed. Ryan Holifield, Jayajit Chakraborty, and Gordon Walker (London: Rutledge, 2018), 1–11; Bruce E. Johansen, *Environmental Racism in the United States and Canada: Seeking Justice and Sustainability* (Santa Barbara: ABC-CLIO,

2020); Edwin Jurriëns, *The Art of Environmental Activism in Indonesia* (London: Routledge, 2023); Fabiana Li, *Unearthing Conflict: Corporate Mining, Activism, and Expertise in Peru* (Durham, NC: Duke University Press, 2015); Mikaela Luttrell-Rowland, *Political Children: Violence, Labor, and Rights in Peru* (Stanford: Stanford University Press, 2023); Devendraraj Madhanagopal et al., eds., *Environment, Climate, and Social Justice: Perspectives and Practices from the Global South* (Berlin: Springer, 2022); Juan Martínez-Alier, "Environmental Justice (Local and Global)," *Capitalism Nature Socialism* 8, no. 1 (1997): 91–107; Michael Mascarenhas, *Lessons in Environmental Justice: From Civil Rights to Black Lives Matter and Idle No More* (Los Angeles: SAGE, 2020); Char Miller and Jeff Crane, *The Nature of Hope: Grassroots Organizing, Environmental Justice, and Political Change* (Louisville: University Press of Colorado, 2019); Rob Nixon, *Slow Violence and the Environmentalism of the Poor* (Cambridge, MA: Harvard University Press, 2011); Stacia Ryder et al., eds., *Environmental Justice in the Anthropocene: From (Un)Just Presents to Just Futures* (London: Routledge, 2021); David Schlosberg, *Defining Environmental Justice: Theories, Movements, and Nature* (Oxford: Oxford University Press, 2009); Dorceta E. Taylor, *Toxic Communities: Environmental Racism, Industrial Pollution, and Residential Mobility* (New York: New York University Press, 2014); Leah Temper, Daniela del Bene, and Joan Martínez-Alier, "Mapping the Frontiers and Front Lines of Global Environmental Justice: The EJAtlas," *Journal of Political Ecology* 22 (2015): 255–78; Gordon Walker, *Environmental Justice: Concepts, Evidence and Politics* (London: Routledge, 2012); Laura Westra and Bill E. Lawson, eds., *Faces of Environmental Racism: Confronting Issues of Global Justice* (Lanham: Rowman & Littlefield, 2001).

2. Sophie Chao and Danielle Celermajer, "Introduction: Multispecies Justice," *Cultural Politics* 19, no. 1 (2023): 2.

3. Chao and Celermajer, "Introduction," 2; Michael Wilson Becerril, *Resisting Extractivism: Peruvian Gold, Everyday Violence, and the Politics of Attention* (Nashville: Vanderbilt University Press, 2021); Christine J. Winter, *Subjects of Intergenerational Justice: Indigenous Philosophy, the Environment and Relationships* (London: Routledge, 2022).

4. Achille Mbembe, "Necropolitics," trans. Libby Meintjes, *Public Culture* 15, no. 1 (2003): 11.

5. E. M. Dijkstra and M. J. Goedhart, "Development and Validation of the ACSI: Measuring Students' Science Attitudes, Pro-environmental Behaviour, Climate Change Attitudes and Knowledge," *Environmental Education Research* 18, no. 6 (2012): 737.

6. "Environmental Justice! How Can We Create Environments That Are Healthy for Everyone?," Smithsonian Science Education Center, © 2022, https://ssec.si.edu/

environmental-justice.

7. Kuk-Kyoung Moon, Seo-Hee Lee, and Seo-Yeon Jeong, "Examining the Relationship between Individualism and Pro-Environmental Behavior: The Moderating Role of Social Cohesion," *Behavioral Sciences* 13, no. 8 (2023): 2.

8. Clare Echterling, "How to Save the World and Other Lessons from Children's Environmental Literature," *Children's Literature in Education* 47 (2016): 286–89.

9. Chao and Celermajer, "Introduction," 3.

10. Jennifer Wenzel, *The Disposition of Nature: Environmental Crisis and World Literature* (New York: Fordham University Press, 2020), 4–5.

11. Wenzel, *The Disposition of Nature*, 5.

12. See, for example, "Indigenous Voices Need to Lead Australia's Response to the Climate Crisis," University of Sydney, August 11, 2023, https://www.sydney.edu.au/news-opinion/news/2023/08/11/indigenous-voices-need-to-lead-australias-response-to-the-climat.html; Veronica Matthews et al., "Justice, Culture, and Relationships: Australian Indigenous Prescription for Planetary Health," *Science* 381, no. 6658 (2023): 636–41.

13. Eben Kirksey and Sophie Chao, "Introduction: Who Benefits from Multispecies Justice?," in *The Promise of Multispecies Justice*, ed. Sophie Chao, Karin Bolender, and Eben Kirksey (Durham, NC: Duke University Press, 2022), 14–15.

14. See also Tatiana Konrad, ed., *Disability, the Environment, and Colonialism* (Philadelphia: Temple University Press, 2024).

15. Karen Bell, "Diversity and Inclusion in Environmentalism," in *Diversity and Inclusion in Environmentalism*, ed. Karen Bell (London: Routledge, 2021), 3.

16. Paul Jobin, Ming-sho Ho, and Hsin-Huang Michael Hsiao, "Environmental Movements and Politics of the Asian Anthropocene: An Introduction," in *Environmental Movements and Politics of the Asian Anthropocene*, ed. Paul Jobin, Ming-sho Ho, and Hsin-Huang Michael Hsiao (Singapore: ISEAS—Yusof Ishak Institute, 2021), 3–4.

17. Manuel Arias-Maldonado, "Politics in the Anthropocene," in *Altered Earth: Getting the Anthropocene Right*, ed. Julia Adeney Thomas (Cambridge: Cambridge University Press, 2022), 160; see also Tatiana Konrad, *Climate Change Fiction and Ecocultural Crisis: The Industrial Revolution to the Present* (Reno: University of Nevada Press, 2024).

18. Sumudu Atapattu and Carmen G. Gonzalez, "The North-South Divide in International Environmental Law: Framing the Issues," in *International Environmental Law and the Global South*, ed. Shawkat Alam, Sumudu Atapattu, Carmen G. Gonzalez, and Jona Razzaque (Cambridge: Cambridge University Press, 2015), 5.

19. Kirksey and Chao, "Introduction," 2.

20. Kirksey and Chao, "Introduction," 3.

21. Sophie Chao and Eben Kirksey, "Glossary: Species of Justice," in Chao, Bolender, and Kirksey, *The Promise of Multispecies Justice*, 23.

22. Thom Davis and Alice Mah, "Introduction: Tackling Environmental Injustice in a Post-Truth Age," in *Toxic Truths: Environmental Justice and Citizen Science in a Post-Truth Age*, ed. Thom Davies and Alice Mah (Manchester: Manchester University Press, 2020), 1.

Part 1

Air, Breath, and (In)Visibility of Environmental Racism

Atmoracism

Air and Precarities of Health, Environment, and Race

Tatiana Konrad

Aerobic beings require oxygen to live. Air provides humans and other creatures with this vital element. Interacting with air continuously through breathing, aerobic beings are oftentimes unaware of the profound significance that this invisible substance has for us all. Having access to, being enveloped by, and existing in air, we depend on it without ever being cognizant of this complex relationship. What changes, then, if this symbiosis is jeopardized or sabotaged? While access to (clean) air should be a fundamental human right, the stories of those who live in polluted areas, those who got infected with the airborne coronavirus having breathed in the viral droplets, or those who were suffocated by racist police officers illustrate the opposite: namely, how air becomes segmented into healthy and unhealthy spaces, and how it can be weaponized against Black, Indigenous, People of Color (BIPOC) and poor communities to sustain abuse, oppression, and extermination.

In what follows, I engage with the peculiar type of racism that BIPOC communities face, coining the concept of *atmoracism* to describe oppression through the medium of air. Working at the intersection of health, environment, and race, I reveal how exposure to air pollution, the airborne COVID-19 virus, and interference with the breathing of Black bodies, as examples of atmoracism, illustrate a fundamentally venomous type of racism that is not simply an instrument of subjugation but the methodical, continuous, and unstoppable mutilation, destruction, and extermination of people of color. The invisibility of air dangerously veils atmoracism and keeps BIPOC communities literally suffocating: breathing toxic air, fighting to breathe through lungs damaged by the airborne virus, or facing police brutality. Pivotally, this essay does not suggest that atmoracism is the only or most concerning type of racism that must be addressed, but rather sheds light on yet another form of structural oppression that, due to its connection to the "invisible" air, might remain unnoticed and neglected by scholars. The ubiquitous nature of atmoracism directed at BIPOC communities is alarming. Seeing air and recognizing its overtly material function can help identify cases of atmoracism and, through this, fight for racial equality and social justice. This type of social justice is inseparable from environmental justice, health justice, and racial justice. This essay thus examines the precarious conditions in which BIPOC communities exist, theorizing atmoracism through the lens of environmental–health–racial justice.

Environmental humanities and media studies scholars emphasize the unique role of documentaries in communicating environmental problems, spreading eco-awareness, and provoking pro-environmental action. John A. Duvall distinguishes between environmental documentaries that address larger topics, such as the impact of humanity on the planet as a whole, and films that explore the local and concrete ways in which environmental devastation occurs, from pollution to species extinction and beyond. In doing so, documentary films, according to Duvall, work in different ways, from serving as a pure source of information to emphasizing specific actions that must be taken in order to minimize or eliminate the effects of a crisis.[1] In either way, the environmental documentary is a politically and socially motivated cinematic genre that, as JoAnn Myer Valenti observes, "gives environmental information a megaphone."[2] Recognizing the important role and unique function of documentaries in promoting environmental justice, this essay draws on PBS's *Hindsight* (2021): six short documentaries that, through the lens of air, focus on racism, the COVID-19 pandemic, and food scarcity, among other problems. Via the analysis of these films, the essay demonstrates how BIPOC use air space to fight oppression, a phenomenon that I term here as *atmoactivism*.

I Can't Breathe: Theorizing Atmoracism through Air

BIPOC communities, among whom are Black Americans, have a complex relationship with air—largely due to the abuse, exploitation, and oppression that they experienced from white colonizers and later, in postcolonial and postslavery times, the dominating white majority that did not and does not welcome BIPOC as equal members of society. The right to breathe clean, free air is undermined by racist and colonial ideologies that treat BIPOC individuals merely as expendable bodies for labor. Much of this oppression has been implemented through air, revealing an even deeper complexity and a distinct type of racism, which I define as atmoracism. This section engages with different instances of racial oppression that were exercised and sustained through air, thus outlining the long, complex, and oftentimes invisible history of racism built on controlling access to this vital element.

Michael Eric Dyson writes about breathing as "biological and metaphoric, both literal and symbolic" for Black people.[3] For African Americans, breathing is not a natural, unnoticeable, thoughtless process of inhaling and exhaling, as is essentially the case for every human who does not have health-related respiratory problems. Breathing, for Black people, is quite literally a form of struggle. Dyson writes about "the air of freedom" that Black people were deprived of once enslaved, as well as of a much more complex deprivation that occurred through slavery:

> When we were kidnaped from our native haunts, we were denied more than the African air we breathed. We were also deprived of the oxygen of opportunities to deepen our ties to our cruelly estranged motherland. And we had withheld from us the nitrogen of nourishment from the land and limbs of our kin and loved ones.[4]

Air and kinship—to land, ancestors, the environment—are tightly connected. Air formulates those multiple instances of kinship; but it is also an element to which one can be kin. And while this segmented view of air is rather problematic in questions related to environmental health, as I will reveal further in this essay, it is helpful in this context to realize that despite its alleged invisibility and imperceptibility, air is an essential part of the larger environment in which humans exist. The connection to home, and the feeling of being homesick when away from home, is conditioned not only through one's relationship with other humans, including relatives, friends, and dwelling place, but also through the environment, the climate of that territory, the weather that is typical of that place—all of which are tightly connected and perceived, among other things, through air. Keeping in mind this relationship among the air, the environment, and home, the deprivation of these three things through enslavement—kidnapping Africans and transporting them to the New World—can be viewed as a form of atmoracism.

Air also played a dramatic role in the process of transporting Africans to the shores of the Americas. Dyson refers to the miasma that filled the parts of the ships where kidnapped Black people were trapped:

> The air we breathed was polluted by the feces and urine of our fellow captives. We imbibed the stench of vomit from souls who couldn't stomach the sadistic and cruel treatment. We could barely breathe.... Some of us had our breath snatched from us at the end of a blunt instrument or by the barrel of a weapon. Some of us stopped breathing as we jumped, or were pushed, overboard.[5]

Here, the relationship between the deprivation of slavery and air is vividly illustrated through breathing. Specifically, the air is unhealthy, and breathing, while necessary to survive, also ruins one's health through inhaling different pathogens. Yet even such polluted air was not a given for the enslaved people, as the history of transporting kidnapped Africans reveals, for access to it was controlled and mediated by the enslaver.

African Americans have been deprived of clean air since the moment their ancestors were brought on the ships; this denial continues today through air pollution. Dyson laments that there is a new method of oppressing and eradicating Black people, namely through "the relentless spray of contaminants" that "pollute[s] the emancipated air around us."[6] Toxic air has been an alarming concern for BIPOC communities, who are literally suffocating on polluted air on a daily basis. Recent documentaries, including *The Sacrifice Zone* (2020), *Unbreathable: The Fight for Healthy Air* (2020), and *Conviction* (2021), effectively illustrate this problem. Magdalena Górska claims that through the relationship between breathing and the quality of air, "the dynamics of geopolitical economic and (neo)colonialist power relations" become apparent. Specifically, "political, social and economic distribution and maintenance of privilege and lack thereof, and power that materializes not only in (un)breathable and (non)toxic air but also in political, social and ethical matters such as whose lives are breathable and whose loss of breath is grievable" are all crucial questions that can be addressed through air.[7] Concentration or relocation of pollutants into the areas primarily populated by BIPOC and poor communities overtly demonstrates whose lives matter and whose do not.

Environmental racism, a term coined in the 1980s, recognized the effects of pollution and other environmental inequalities as systemic oppression. In 1987, Benjamin Chavis explained environmental racism as

> racial discrimination in environmental policy-making and enforcement of regulations and laws, the deliberate targeting of communities of color for toxic waste facilities, the official sanctioning of the presence of life-threatening poisons and pollutants for communities of color, and the history of excluding people of color from leadership of the environmental movement.[8]

Before Chavis, however, Robert D. Bullard described the problem of environmental racism in the following way: "People of color are subjected to a disproportionately large number of health and environmental risks in their neighborhoods . . . and on their jobs."[9] Today, environmental racism remains an urgent problem, not only revealing social, political, economic, environmental, and medical inequalities, but also literally destroying the health and lives of BIPOC communities. Stacy Alaimo notes: "Race . . . has been well documented as the single most important factor in the placement of toxic waste sites in the United States."[10] And there are multiple examples that would prove this, including the infamous story of Uniontown, Alabama:

> Few places define environmental racism as precisely as Uniontown, Alabama, 30 miles west of Selma, home of the Arrowhead Landfill, into which, for 18 months beginning in July 2009, hundreds of railcars a day poured more than 4 million tons of coal ash laced with arsenic, lead, and several other heavy metals. Uniontown's 2,400 residents are more than 90 percent black. The ash has come from a coal-burning power plant in Roane County, Tennessee, 300 miles north. The population of Roane is roughly 90 percent white.[11]

Instances like this one distinctly speak about racism, environmental racism, but also, more specifically, atmoracism, directed at BIPOC communities, and illustrate how, when fighting for racial equality, it is crucial to keep in mind the environment, including air, as sites of discrimination. In line with Alaimo, I emphasize that "social injustice is inseparable from physical environments," for "human bodies, human health, and human rights are interconnected with the material, often toxic flows of particular places."[12] Recognizing this, and better understanding these complex linkages, can help outline the ways in which "civil rights, affirmative action, and identity politics models of social justice" have to be transformed to account for both the lingering effects of historical instances of environmental racism and its current, daily occurrences.[13]

BIPOC communities are exposed to multiple kinds of toxicity, including, for example, pollution that emerges from farms and industrial sites, but also air, water, and land pollution, more broadly.[14] The mosaic nature of pollution, which harms some communities but not others, can be approached through what Dorceta E. Taylor terms as "residential segregation"—"a pervasive feature of American society"—that "is related to exposure to environmental hazards": even financial stability does not guard BIPOC from segregation and pollution.[15] Through the visual mapping of such segregated communities, it becomes apparent that BIPOC are ecological outcasts. Sarah Jaquette Ray's concept of "ecological others" is particularly helpful to understand the degree to which BIPOC individuals are excluded from the landscape of health and environmental well-being. Jaquette Ray claims: "Unlike ecological subjects, whose aim it is to save the world from ecological crisis, ecological others are often those from whose poor decisions and reckless activities the world ostensibly needs to be saved."[16] She draws on Susan Kollin, who explains how "ecological others" emerge: "The loss of nature experienced by Euro-Americans often becomes directed toward the racial Other, who in turn is made responsible for that loss, becoming a target of environmentalism's denigration and blame."[17] Through racial othering, environmental racism operates on different levels to sustain environmental degradation and racial inequality.

Atmoracism is a form of environmental-racial injustice that can be traced back to the times of slavery, when, as scholars energetically insist today, oppression occurred both at the hand of men and through the environment. The enslaver formulated a peculiar attitude to the land and the enslaved populations, designating both as "disposable."[18] Today, in the twenty-first century, "systemic racism continues to intersect with health disparities, environmental disparities, and police violence to deem certain groups disposable."[19] All these types of oppression are directly connected to the air; they aim at harming or destroying BIPOC health.[20]

In the time of COVID-19, the existing instances of atmoracism are amplified through the airborne pandemic and its disproportionate harm to BIPOC communities. BIPOC health has generally been targeted by systems of oppression to further subjugate these individuals, as, for example, the documentaries *They're Trying to Kill Us* (2021) and *Why Is Covid Killing People of Color?* (2021) reveal. The attack on and vulnerability of minority health became apparent again once COVID-19 hit. "When America has a cold, Black America has pneumonia," writes Dyson.[21] And this is exactly what has been happening during the pandemic. Dyson considers racial inequality through COVID-19 metaphors:

> No masks could hide the fear and hate of Blackness nor stop the spread of viral anti-Blackness. No volume of hand sanitizer could cleanse the grime of diseased Blackness from the hands or knees of loathsome cops. No study of viral ontology could relieve the ontological spite that many cops feel for Black being. No determination to flee the clutches of angry cops could escape the contact trace of bigoted policing. No amount of social distancing—a racially colored idea that sought, more than a century ago, to measure the amount of space that white folk would maintain from Black people in their everyday existence—could keep cops from brutalizing and killing the bodies of Blacks.[22]

The pandemic has certainly helped reveal the endemic nature of racism in the United States, by making the vulnerability and disposability of Black bodies visible. Black people had a 3:1 ratio of death compared to white people. Yet the pandemic was not the only killer of the Black community; racism and anti-Black violence continued, with over two hundred African Americans falling victim to police brutality just in February and March 2020.[23]

Studies reveal that over thirty thousand people die in the United States each year because of air pollution.[24] The coronavirus disease and air pollution are a deadly combination. Communities of color disproportionately suffer from heart and lung illnesses as well as COVID-19; these same communities live in places with high air pollution. Research shows that people with coronavirus are more likely to die if they are also exposed to air pollution.[25] Importantly, while the particles of air pollution cannot carry the virus, scientists confirm that the impact of pollution on human bodies results in enhanced "susceptibility" to the airborne virus.[26] One vivid example is "Cancer Alley"—air-polluted LaPlace, Louisiana, where more than half of the population are African Americans. LaPlace is known to have had the highest death rate from the coronavirus per capita in the whole United States in early 2020.[27] Race has literally become a scale against which one can forecast and measure pollution.[28]

There are thus dangerous dynamics between pollution and the coronavirus that can be read through air. In combination, air pollution and the airborne virus destroy respiratory systems, but they also pollute places, turning specific areas into zones of "bad air." There is, however, a major problem with zoning polluted air. On the one hand, areas with high levels of pollution can be clearly mapped: the locations of identified polluting industries are known, and the harm that they do to environments and their inhabitants can be assessed. Through this simple formula, one can clearly observe the damage caused to BIPOC communities. On the other hand, this approach can falsely suggest that pollution and death can be isolated within some invisible borders—as invisible as the air itself. This is, of course, not accurate. The concentration of polluting particles can be higher in some places than others, yet the air cannot be segmented. Undoubtedly, disproportionate harm accrues to BIPOC communities through various forms of atmoracism, yet neither does the rest of the planet remain untouched. Realizing this can help work against atmoracism and toward the right for clean air for oppressed communities, as well as for all other human and more-than-human inhabitants of this planet.

Atmoracism is tightly intertwined with breathing, for it is through breathing, or lack thereof, that one can recognize suffocation as a racialized process. Górska emphasizes the politicization of breathing through its dual nature, namely that breathing is an ordinary, and even habitual, action that is also different in every instance, thus revealing its contested nature:

> In its persistent commonality and constant differentiation, breathing . . . can inspire diverse analyses of relational natural and cultural, material and social scapes that are oxygenated across diverse spaces, times, geopolitical relations, ecosystems, industries and urbanization while being situated in their phenomenal specificities. It becomes an enactment of movement and circulation within and across (human and nonhuman) bodies, spaces, species and cultures. Also, multiple forms of breathing (such as with

technologies or with different kinds of air and the dust or pollution it contains) have the power to articulate how societal power relations materialize in and through, and are enacted by, bodies.[29]

BIPOC breath is controlled. Through breath, air becomes controlled, too, resulting in forms of atmoracism. The ongoing suffocation of BIPOC communities intermingles with the interruption of breath and ultimate death. While there is a temptation to speak about this interruption of breath in a figurative sense, in fact, it happens all the time literally. Dyson, for example, connects enslavement to air and foregrounds the impact that bondage had on Black people through this element. He claims that once transported to the New World, enslaved individuals "could barely breathe free air for centuries afterward."[30] The unbreathability of slavery, though it lends itself to metaphorical descriptions of the lack of freedom, also quite literally chronicles the suffocation and death of enslaved people. From the lack of physical space and contamination in the holds of slave ships, to the toxicity that surrounds and harms Black bodies in the post-Reconstruction United States, including through coal mining, toxic waste, air pollution, and other hazardous conditions, unbreathable air has been enveloping BIPOC communities, destroying their health and causing premature death. Dyson's affective metaphor of the lack of breath—"From the start of our forced intimacy with North America, Black folk have been trying to breathe air that is free of the pollution of captivity, of coerced transport, of enslavement, of white supremacy, of social inequality and perennial second-class citizenship"—therefore, carefully documents the existing harm, violence, and crime directed at BIPOC communities that construct atmoracism and result in real cases of deterioration of human health and death.[31]

Today, racial inequality and the fading of BIPOC breath are further reinforced through the airborne pandemic and police brutality, both of which are forms of atmoracism. The coronavirus pandemic has not only made inequality visible, but also emphasized how colonial structures, ideologies, and policies continue to operate on various levels to subjugate BIPOC communities. Health as such has been formulated through colonialism.[32] Colonialism reimagined Black and brown bodies as tools for white people's prosperity, as obstacles to white people's domination, and as replaceable bodies grown and used for servitude. The ongoing pandemic has revealed colonial understandings of BIPOC as bodies that suffer and do not matter to the white majority. The coronavirus pandemic, and specifically the disproportionate harm that it has had on BIPOC communities, has uncovered and contributed to atmoracism.

But the absence of breath, and the absence of air, is caused not only by polluting particles or viruses but also by humans. The infamous cases of police brutality—committed against Eric Garner in 2014 and George Floyd in 2020— illustrate how atmoracism can be directly performed by human agents. "I can't breathe," the words uttered by both Garner and Floyd, denote the actual, physical effects of the violence on their bodies. But they also illuminate the control over Black breathing exercised by the white police officers, as well as the precarity of Black existence through the air of which these Black men were deprived. There is an axiological correlation between air and respiration.[33] The interference with the natural breathing process as well as suspension of one's access to air are simultaneously forms of social and environmental injustice, for these processes endanger the human rights to existence and to a clean and healthy environment. Air is inseparable from the ongoing crises. Scholars observe the interlinking relationship among the COVID-19 pandemic, police brutality and racism more broadly, and the environmental crisis.[34] The words "I can't breathe" equally describe the suffering that BIPOC experience due to air pollution, the airborne pandemic, and police brutality.[35] All of these processes have become (or initially were) racialized through the interference with air to control BIPOC breath.

There are multiple ways to read air. Nerea Calvillo, for example, underscores its myriad metaphorical meanings, viewing it as "a space, an object, a threat, a myth, a weapon, a common."[36] In the context of anti-Black violence, air is a weapon, and atmoracism an act of harm. One of the most influential readings of air as a weapon has been provided by Peter Sloterdijk, who analyzed the use of gas during World War I as an attack through the air. Sloterdijk argues: "By using violence against the very air that groups breathe, the human being's immediate atmospheric envelope is transformed into something whose intactness or non-intactness is henceforth a question." Sloterdijk's understanding of what it means to attack a breathing body through air as a form of "atmoterrorism" directly connects to the ongoing forms of oppression through the air.[37] Sloterdijk classifies "atmoterrorism" as a military action, "environmental warfare" through which an "attack on the enemy's primary, ecologically dependent vital functions" is exercised.[38] There is, however, a fundamental difference between "atmoterrorism" and atmoracism. While both emphasize the toxicity of actions against breath, foregrounding the breathing body and its vital dependence on air, atmoracism is a much more precise, deliberate, and destructive violence that is spread globally and targets selectively. Rob Nixon's concept of "slow violence," as "a violence that occurs gradually and out of sight, a violence of delayed destruction that is dispersed across time and space, an attritional violence that is typically not viewed as violence at all," accurately describes the way atmoracism operates.[39] The invisibility of "slow violence" also directly parallels the alleged invisibility of air, foregrounding how the crimes against BIPOC communities have been veiled and trivialized through selective vision. It is telling that police brutality has proved to be the most recognizable form of violence among all the other forms of atmoracism, because it is easily seen as violence as such, that is, "an event or action that is immediate in time, ... and as erupting into instant sensational visibility."[40] Directly conceived of as a form of anti-Black violence, police brutality has sparked responses from Black communities and antiracists worldwide.

Air is not safe.[41] It is polluted with toxins, filled with viruses, and controlled by select groups and individuals (consider, for example, how air is polluted by industries that define air spaces as healthy or unhealthy). Air is particularly unsafe for BIPOC individuals. Through air, white people control BIPOC power and positionality in the world. For breathing bodies, air is a vehicle through which to interact with the world via breath, breath being "a mode of relating to the world, engaging with others, objects, environments and technologies."[42] Air communicates relationships, positions, deprivations, abuse, and destruction, among other things. Air tells "airstories," that is, interactions with air and breath, formulating "collective awareness of ecopolitical intra-actions and co-constitution with air."[43] When approached through the lens of atmoracism, racial injustice becomes a complex entanglement of environmental, health, political, and economic factors that are used against BIPOC communities. It is thus impossible to address social justice without environmental justice. In fact, Bryan K. Bullock argues: "The environmental justice movement may be the most important civil rights issue of the 21st century. It bridges the gap between environmental concerns, civil rights and human rights."[44] But racial justice equally contributes to the restoration of environmental justice. At the same time, when approached through the pandemic, the environmental crisis becomes an even more acute problem.[45] Simultaneously, the pandemic makes racial inequality more apparent. One way to address the linked problems of racism, environmental degradation, and the pandemic is through air. In their introduction to *The Bloomsbury Handbook to the*

Medical-Environmental Humanities, Scott Slovic, Swarnalatha Rangarajan, and Vidya Sarveswaran view health, in light of the coronavirus pandemic, as "one" complex entity: "There is no 'human health' and 'environmental health.' There is only one health—and the absence of health. There is only precarity, experienced admittedly in variable degrees according to just and unjust social and economic systems, but an overarching and fundamental precarity that encompasses all of us, regardless of nationality and ethnicity and species."[46] Thinking about the unity of health is productive in combating atmoracism, for this approach calls for medical equality, environmental justice, and the right to breathe clean air. Combating atmoracism means restoring social, racial, and environmental justice, as well as recognizing the mutual coherence of its components and the impossibility of solving one without addressing another.

PBS's *Hindsight*: BIPOC Communities and Atmoactivism

While there are multiple documentaries that overtly engage with atmoracism (some of which are mentioned above), here I intentionally focus on the films that work against atmoracism, through atmoactivism, to foreground the struggles of BIPOC communities to end racism that is experienced through the air.

PBS's *Hindsight* includes six short documentary films. Ranging from ten to fifteen minutes, these documentaries are snapshots of BIPOC people's lives during the coronavirus pandemic in 2020. Dilsey Davis's *Now Let Us Sing*, Anissa Latham's *Missing Magic*, Kiyoko McCrae's *We Stay in the House*, Zac Manuel's *This Body*, and Amman Abbasi's *Udaan* (*Soar*) focus on the U.S. South, including locations in North Carolina, Alabama, Louisiana, and Arkansas, whereas Arleen Cruz-Alicea's *Comida pa' los Pobres* (*Food for the Poor*) takes place in Puerto Rico. The documentaries explore racial inequality through issues of Black (single) motherhood, police brutality, and the life of BIPOC communities at the start of COVID-19. Air appears in these stories visually, but also through the personal stories that explicate BIPOC suffocation in a world filled with injustice, which has only further intensified during the pandemic.

Now Let Us Sing focuses on the interracial One Human Family Choir in Durham, North Carolina, whose members are devoted to racial equality and peaceful existence among all races. Dilsey Davis is the protagonist in this film;

through her perspective, the film explores how the members of the choir have been coping with the pandemic and police brutality—two forces that threaten BIPOC lives. Through Dilsey, the viewer hears the choir's decision to adjust to the online format and continue to meet and sing via Zoom. By articulating the racism that Dilsey faced as a child and current racist realities, largely explored through the case of George Floyd and other instances of police violence, the documentary showcases the choir's important work to promote justice and equality through open conversations about race. While the choir has played a significant role in the lives of its members, the pandemic, it seems, has deprived them of the safe, activist, equal space that they have created together.

The transition to the online format proves to be difficult at first; unstable Internet connections, sound issues, and multiple other problems have transformed the singing experience into something entirely new. Yet the singers persevere, as they identify ways that help them continue to exist as a community and do what they like best—sing. Dilsey describes singing as a transformative experience, as an opportunity to escape the racist reality that she has to deal with on a daily basis and be alone with herself. When she is practicing, she finds herself "in this space, and nothing else matters."[47] The opportunity to continue singing despite the lockdown is thus a way to maintain the shelter created through the choir. Yet singing online is a peculiar process; everyone must be muted because of the connection and sound issues. "It's like singing into a vacuum." Though the community remains as a whole, engagement among its members is just not the same as when they met in person. Dilsey's comments depict song as a function of air, illustrating how, through air, collectivity and subjectivity can remain intact but also be disrupted. Singing is like being in a bubble, an air space to which no one has access and in which racial harmony can be restored. This is complicated when singing as a choir, as the singers must mute themselves in order to hear each other. The vacuum to which the documentary refers thus does not necessarily connote the futility of seeking racial justice through song, but rather making a space in which everyone's voice can be heard. Dilsey emphasizes the power of song over individuals and communities. Saying, "I have to continue to sing. I have to continue to have a voice," Dilsey not only acknowledges the impact singing has had on her life but also foregrounds the political value of singing; that is, through singing, messages can be communicated, individuals get an opportunity to speak and be heard, and the voices of marginalized people become audible, all of which are instruments to fight racism.

While *Now Let Us Sing* mediates the tragedy of the pandemic through the choir, *We Stay in the House* addresses disease and loss directly, opening with the images of a Latina woman whose health deteriorates dramatically as she becomes infected with COVID-19. She shares with the viewer how anxious she is to even walk to the bathroom, "'cause I can't breathe [pausing and exhaling loudly several times] without oxygen."⁴⁸ The woman survives, yet her scary experience transforms her and other BIPOC women who know her. This film works at the intersection of gender and race, focusing on BIPOC women, mothering, and work-life balance (or lack thereof) in the lockdown, explicating the particular problems that women faced once COVID-19 hit. These experiences communicate another way that COVID-19 suffocated BIPOC women; although not infected, they can hardly breathe taking care of their children, working several jobs to feed their families, losing jobs and searching for new ones, and trying to find temporary spaces where they can be productive as mothers and employees.

Policing BIPOC bodies—an issue that is addressed in *Now Let Us Sing*—is further explored in *Missing Magic* and *Comida pa' los Pobres* (*Food for the Poor*) that thoroughly examine brutality and injustice at the hands of the police. These documentaries illustrate how racial profiling is an everyday reality for BIPOC communities; demonstrate how their lives, words, and actions are surveilled by the state; and illuminate their subordinate position in a society that privileges white people and promotes white supremacy. Just like *Now Let Us Sing*, *Missing Magic* tackles the infamous murder of George Floyd through the eyes of a young African American poet who uses spoken-word poetry to express his indignation with racial inequality faced by Black people. Saying out loud the names of those murdered by police officers means keeping the memory of those people alive, making the injustice audible, and fighting for equality through one's own voice. Engaging with breath and air, American ecologist and philosopher David Abram observes the unique power of air as a function of speech:

> It's very hard to speak on the inbreath but on the outbreath, that's what we shape with our mouth and our tongue and we speak. And so it is assumed that it is the breath that is carrying my words to your ears or your words to my ears. So the air is the implicit intermediary in all communication. *The air is the very medium of meaning.* It is the very place of the spirits. It is the place where your ancestors [*sic*] voices linger after they've passed on, the air.⁴⁹

Abram's characterization of air as a "medium of meaning" helps to contextualize the singing depicted in *Now Let Us Sing* and spoken-word poetry in *Missing Magic* as forms of political power and meaning-making. Voicing injustice is a way to fight injustice. This exact message is communicated in *Comida pa' los Pobres (Food for the Poor)*, the final documentary in the *Hindsight* series, which tells the story of young Puerto Rican activist Giovanni, who helps the poor by providing food. This mission is largely complicated by the pandemic, which makes the food so much more expensive. Giovanni's personal experience with hunger, the stories that he witnesses on the island, and the ongoing colonial devastation that keeps the territory in a perpetual crisis prompt him to rise up and speak for those whose voices have become silent. Giovanni is soon arrested, yet in the cell, he chooses to sing. Banging with a plastic bottle filled with frozen water at the cell's bars, Giovanni sings: "What I want is food for the poor, food for the poor."[50] Giovanni explains, "It's working! Whether the cops want to or not they're listening to what I'm singing. And I went on, and on, and on. I wanted to get in their heads. I wanted these cops to go to sleep that night hearing my song. That was my intention, full on. These people are going to dream about me." Engaging with the air through speaking—to communicate ideas and make the voices of those who are oppressed audible—is a form of political activism that works at the intersection of social-environmental justice.

The theme of policing BIPOC bodies is further developed in *Udaan (Soar)* and *This Body*. The former tells the story of young Pakistani woman Baneen who moves to the United States to get an education. Baneen travels to the United States with her mother, who will help her adjust to her new life abroad. Yet on the day the two women land in the United States, the insurrection at the Capitol occurs. After a long interrogation, Baneen's mother is sent back to Pakistan because her visa expired during the pandemic, while Baneen is allowed to enter the country. *This Body*, in turn, explores the complex relationship between Black Americans and medicine. Aware of the high COVID-19 death rates among African Americans, Sydney Hall decides to participate in the vaccine trial, hoping to help her community. The documentary makes references to the history of medical racism and abuse, drawing parallels between police brutality and medical racism. Sydney's decision is a form of activism—pro-life in general, and pro–Black life in particular. It is a powerful way to oppose the racist system and to dismantle medical racism, while remembering its history and ongoing impact on BIPOC communities. Sydney emphasizes that she is aware that the vaccine is experimental, but she foregrounds the larger goal that her action can achieve. There are two options: "Am I just gonna be afraid and just let people keep dying? Or, am I gonna put my fears aside for a second?"[51] Sydney's decision is a form of activism against medical racism, for she both helps save (Black) lives and invites the Black community to move beyond the racist historical past and toward a just present. Both documentaries criticize the vision of BIPOC bodies as "easily disposable," manipulated, surveilled, or controlled otherwise.

Through these six documentaries, *Hindsight* explores the ways racism (including atmoracism) permeates BIPOC lives. From police brutality to medical racism and beyond, the films illustrate oppression and the struggle to live, to breathe. In a unique way, the documentaries show how BIPOC communities' engagement with air can be used as a weapon to fight racism—through the sung and spoken word, making injustice audible, creating new meanings, and restoring equality. *Hindsight*'s atmoactivism is an important form of storytelling that works at the intersection of political–social–environmental justice, merging racial justice and air.

Conclusion

Breathing is an essential, vital function of aerobic bodies. Breathing can be understood as the lungs' engagement with air: "The diaphragm contracts. It drops. A vacuum appears in the chest cavity, which allows the lungs to expand with air. While the lungs are surfeit with air, oxygen passes through thin membranes in the alveoli to bond with haemoglobin, which, in turn, releases its load of carbon dioxide."[52] BIPOC communities are, quite literally, deprived of their right to breathe: to breathe clean air, to breathe through healthy lungs, to breath at all. Control over breathing is largely exercised through air as such: air pollution, the airborne coronavirus pandemic (including the measures taken during the pandemic that sustained precarious health conditions), and the physical obstruction of Black bodies' air passageways by racist police officers are instances of racial injustice that overtly overlap with medical and environmental racism.

The recognition of specific types of structural racism such as atmoracism helps reveal the complexity of racism as it permeates multiples spheres, worsening the lives, health, and living conditions of BIPOC communities. The relationship between racism and air is, however, as tight as the relationship between racial justice and air; BIPOC communities use air as a space, medium, and venue through which to communicate their frustrations, fears, and indignation, and ultimately to fight for justice today and in the future. Atmoracism, just as atmoactivism, reveals that this justice affects not only racial but also health and environmental issues, foregrounding the complexity of the crisis in which BIPOC individuals find themselves and the urgency with which this problem must be addressed to save both human lives and the environment at large.

NOTES

1. John A. Duvall, *The Environmental Documentary: Cinema Activism in the 21st Century* (New York: Bloomsbury Academic, 2017), 30.
2. JoAnn Myer Valenti, "When Environmental Documentary Films Are Journalism," in *Routledge Handbook of Environmental Journalism*, ed. David B. Sachsman and JoAnn Myer Valenti (London: Routledge, 2020), 99.
3. Michael Eric Dyson, "Foreword: Breathe Again," in *Religion, Race, and COVID-19*, ed. Stacey M. Floyd-Thomas (New York: New York University Press, 2022), xi.
4. Dyson, "Foreword," xii.
5. Dyson, "Foreword," xii.
6. Dyson, "Foreword," xii.
7. Magdalena Górska, *Breathing Matters: Feminist Intersectional Politics of Vulnerability* (Linköping: Linköping University, 2016), 30.
8. Benjamin Chavis qtd. in Bruce E. Johansen, *Environmental Racism in the United States and Canada: Seeking Justice and Sustainability* (Santa Barbara: ABC-CLIO, 2020), 1.
9. Robert D. Bullard, introduction to *Confronting Environmental Racism: Voices from the Grassroots*, ed. Robert D. Bullard (Boston: South End Press, 1993), 10.
10. Stacy Alaimo, *Bodily Natures: Science, Environment, and the Material Self* (Bloomington: Indiana University Press, 2010), 29.
11. Johansen, *Environmental Racism in the United States and Canada*, 3.
12. Alaimo, *Bodily Natures*, 29, 62.
13. Alaimo, *Bodily Natures*, 62.
14. "Environmental Racism," Food Empowerment Project, January 2022, https://foodispower.

org/environmental-and-global/environmental-racism/.

15. Dorceta E. Taylor, *Toxic Communities: Environmental Racism, Industrial Pollution, and Residential Mobility* (New York: New York University Press, 2014), 148.

16. Sarah Jaquette Ray, *The Ecological Other: Environmental Exclusion in American Culture* (Tucson: University of Arizona Press, 2013), 5.

17. Susan Kollin qtd. in Jaquette Ray, *The Ecological Other*, 5–6.

18. David Silkenat, *Scars on the Land: An Environmental History of Slavery in the American South* (New York: Oxford University Press, 2022), 1.

19. Heather Leigh Ramos, "Slow Violence, Environmental Toxins, and Systemic Racism: Revisiting Lorde's *Cancer Journals* in the Twenty-First Century," in *The Bloomsbury Handbook to the Medical-Environmental Humanities*, ed. Scott Slovic, Swarnalatha Rangarajan, and Vidya Sarveswaran (London: Bloomsbury Academic, 2022), 143.

20. For more on the relationship between structural racism and health, see Dayna Bowen Matthew, *Just Health: Treating Structural Racism to Heal America* (New York: New York University Press, 2022).

21. Dyson, "Foreword," xii.

22. Dyson, "Foreword," xiv.

23. Stacey M. Floyd-Thomas, "Introduction: The Evidence of Things Not Seen: The Twin Viruses of Blind Faith and Color Blindness," in Floyd-Thomas, *Religion, Race, and COVID-19*, 2.

24. Isabelle Chapman and Drew Kann, "For Some Environmentalists, 'I Can't Breathe' Is about More than Police Brutality," CNN, August 4, 2020, https://edition.cnn.com/2020/06/27/us/environmental-racism-explainer-trnd/index.html.

25. Chapman and Kann, "For Some Environmentalists."

26. Lidia Morawska qtd. in Weijie Zhao, "Changes of Air Quality during the Pandemic and Airborne Transmission Issues," *Natural Science Review* 8, no. 3 (2020): 4.

27. Chapman and Kann, "For Some Environmentalists."

28. Robert Bullard in Chapman and Kann, "For Some Environmentalists."

29. Górska, *Breathing Matters*, 30.

30. Dyson, "Foreword," xii.

31. Dyson, "Foreword," xii.

32. Rosemary J. Jolly, "Pandemic Crises: The Anthropocene as Pathogenic Cycle," *ISLE: Interdisciplinary Studies in Literature and Environment* 27, no. 4 (2020): 816.

33. See Tatiana Konrad, ed., *Imagining Air: Cultural Axiology and the Politics of Invisibility* (Exeter: University of Exeter Press, 2023).

34. Robert-Jan Wille, "Keep Focusing on the Air: COVID-19 and the Historical Value of an Atmospheric Sensibility," *Journal for the History of Environment and Society* 5 (2020): 182.

35. See Chapman and Kann, "For Some Environmentalists," and Dyson, "Foreword," xiii.

36. Nerea Calvillo, "Particular Sensibilities," *e-flux Architecture*, October 2018, https://www.e-flux.com/architecture/accumulation/217054/particular-sensibilities/.

37. Peter Sloterdijk, *Terror from the Air*, trans. Amy Patton and Steve Corcoran (Cambridge, MA: MIT Press, 2009), 25.

38. Sloterdijk, *Terror from the Air*, 15.

39. Rob Nixon, *Slow Violence and the Environmentalism of the Poor* (Cambridge, MA: Harvard University Press, 2011), 2.

40. Nixon, *Slow Violence*, 2.

41. On the safety of air, see Gregers Andeersen, "Greening the Sphere: Towards an Eco-Ethics for the Local and Artificial," *Symploke* 21, no. 1–2 (2013): 138.

42. Rebecca Oxley and Andrew Russell, "Interdisciplinary Perspectives on Breath, Body and World," *Body & Society* 26, no. 2 (2020): 4.

43. Greta Gaard, "(Un)storied Air, Breath and Embodiment," *ISLE: Interdisciplinary Studies in Literature and Environment* 29, no. 2 (2022): 300.

44. Bryan K. Bullock, "'Environmental Justice' Obscures Persistent Racism," in *Environmental Racism and Classism*, ed. Anne Cunningham (New York: Greenhaven Publishing, 2017), 83.

45. Bruno Latour, *After Lockdown: A Metamorphosis*, trans. Julie Rose (Cambridge: Polity Press, 2021), 3.

46. Scott Slovic, Swarnalatha Rangarajan, and Vidya Sarveswaran, "Introduction: Toward a Medical-Environmental Humanities. Why Now?," in Slovic, Rangarajan, and Sarveswaran, *The Bloomsbury Handbook to the Medical-Environmental Humanities*, 9.

47. *Now Let Us Sing*, directed by Dilsey Davis, *Hindsight* (2021), PBS, July 23, 2021, https://www.pbs.org/video/now-let-us-sing-00esmd/.

48. *We Stay in the House*, directed by Kiyoko McCrae, *Hindsight* (2021), PBS, July 23, 2021, https://www.pbs.org/video/we-stay-in-the-house-ace6rz/.

49. David Abram, "The Commonwealth of Breath: Climate and Consciousness in a More-than-Human World," Harvard Divinity School, Center for the Study of World Religions, April 9, 2019, https://cswr.hds.harvard.edu/news/2021/05/10/video-commonwealth-breath-climate-and-consciousness-more-human-world (my emphasis).

50. *Comida pa' los Pobres* (*Food for the Poor*), directed by Arleen Cruz-Alicea, *Hindsight* (2021), PBS, July 23, 2021, https://www.pbs.org/video/comida-pa-los-pobres-food-for-the-poor-n7pvba/.

51. *This Body*, directed by Zac Manuel, *Hindsight* (2021), PBS, July 23, 2021, https://www.pbs.org/video/this-body-1csiy4/.

52. Arthur Rose, "Introduction: Reading Breath in Literature," in *Reading Breath in Literature*, ed. Arthur Rose, Stefanie Heine, Naya Tsentourou, Corinne Saunders, and Peter Garratt (Cham: Palgrave Macmillan, 2019), 1.

The Respiratory Politics of Air-Breath

India's Epidemic Intensities

Bishnupriya Ghosh

O ne of the indirect, and paradoxical, effects of the airborne pandemic across the world was the sudden emergence of clean air. As we watched the temporal dissolve from industrial air heavy with particles to a clearing atmosphere on CNN's *World News*, we apprehended clean air precisely when it disappeared visually. One widely circulated photograph of India Gate situated the planet's atmosphere in the recognizable environs of New Delhi (see figure 1). The media form of the dissolve marked the fragility of "breathable air," always under threat, always a state of exception in industrial modernity—that "doomsday machine that we have built for ourselves," as Arundhati Roy put it in the early days of the pandemic.[1] Encoding the elemental medium, the temporal notation made air *legible* as a shared extensive environment even as the historical experience of our present atmospheric degradation induced apprehensions beyond the visual frame. One could intuit the constriction, then expansion, of the lung in sensory perceptions of air densities. Images of this variety became generic during the first months of the COVID lockdowns, organizing air-breath around iconic images such as the Eiffel Tower, the Venice canals, or Rio's Christ the Redeemer. These monumental icons

FIGURE 1. India Gate environs before and after the March 25, 2020, lockdown, CNN Video.
Geospatial World, April 23, 2020.

immediately instituted geopolitical difference while the seriality of photographs rendered each location equivalent to the other. The upshot was the unexpected emergence of universal utopic possibilities, of common breathable futures, amid a global health emergency.

Alongside this genre came others attuned to highly localized pandemic experiences of air. In the case of Delhi, Pulitzer Prize–winning journalist Danish Siddiqui's drone footage of the Seemapuri crematorium provided unerring evidence of a massive escalation of deaths during the manmade disaster that was India's second wave (April–May 2021). That Siddiqui was a highly credentialed photographer (working with the Reuters multimedia team in Delhi) made the drone image credible at a time when the Narendra Modi government refused to acknowledge the scale of devastation.[2] More shocking was the fact that his aerial footage revealed not just the crematorium but its extension into an adjacent empty lot, now cleared of trees and shrubs. As we know, not only makeshift crematoria but also the demand for wood to fuel funeral pyres placed the city's trees at high risk; indeed, at the height of the second wave, the forest department had to issue special permissions for the felling of trees to meet the impossible demand for firewood.[3] Siddiqui's multimedia footage on the Reuters Graphics platform (published in May 2021) captured the loss of vegetation diagrammatically by marking the "before" and "after" to the Seemapuri crematorium extension.[4] His images revealed not only the loss of vegetation, but smoke densities punctuated by glowing fires, searing tongues leaping at the multistoried building next to the cleared ground. Across global media platforms, horrified viewers watching the tragedy unfold apprehended the stench of death become quotidian. Siddiqui's footage was followed by a deluge of crematoria images in global mass media historically particular to India's COVID experiences. They persist as *the* afterimage—that indelible sensory impress that lingers long after the event—as Zahid Chaudhury noted of traumatic images.[5] As such, the crematoria images function as a wake-up call for the death-dealing consequences of damaged air-breath.

From the standpoint of visual culture, air-breath images are the ingredients for a political aesthetic that tends toward environmental justice: at stake is a respiratory politics of air-breath. These air-breath images are heterogeneous in the sense they belong to different image cultures, including climatic media, drone news footage, feature photography, and documentary. Thematically distinct, they are discontinuous in form, platform, and target audience: as such, I characterize them as separate image clusters. Yet a logical relation between them obtains from the strange temporalities of catastrophes, always coming and already here. In what follows, I plot this relation as the formal syntax of a political aesthetic; specifically, a situated political aesthetic for Delhi's COVID experience. Among the image clusters, the crematoria images signaling mass death operate as one of those "powerful cultural symbols," as Ernesto Laclau once argued, that "quilt together" all others.[6] This quilting involves desires and aspirations, investments that lend affective power to the image. And here lies the promise of the media forms that encode the image, as Brian Larkin elaborates in his introduction to a volume on the political aesthetics of infrastructure.[7] Accompanying legible proprieties, what can images induce? I argue that they can orient spectators/viewers toward other times, spaces, and agencies: the haptic sense of particle-heavy air entering the lung in the India Gate image, for instance, is a "sensory apprehension" of air-breath that is not formally represented within the media form of the video footage. As Jacques Rancière might have it, I suggest what might be partitioned away from the image proper, but intuited, felt, and anticipated.[8] The suggestions of the apprehensible are based on historical experiences, some generalizable, others particular to local circumstance. The onslaught of SARS-CoV-2 on the lungs, to take a simple example, makes it difficult to think about air without its intensive articulation as breath: this is a generalizable experience arising from the COVID pandemic. All in all, one of the aims of this essay is to flesh out what is meant by a "political aesthetic" for this pandemic. This effort is spurred by the question: How might we understand a political aesthetic to evaluate the respiratory politics it mobilizes?

The question orients this critical overture toward environmental justice. This essay takes the form of a COVID wake by drawing inspiration from Christina Sharpe's landmark *In the Wake: On Blackness and Being*. Mobilizing the slave ship as a powerful metaphor for thinking about contemporary Black experiences of punishment, regulation, incarceration, and death, Sharpe speaks to the afterlife of slavery. Black death is normative and necessary for "this so-called democracy [the United States]," she surmises; "it is the very ground we walk on."[9] With other scholars of global racial capitalism, she underscores the persistence of global anti-Blackness as the root of premature and preventable Black death become normative, fueled as it is by everything from long-term health-care inequities to extrajuridical murders. Sharpe's call to a wake for the uncounted informs not only my collection, collation, assembly, and circulation of image clusters from India's COVID experience, but my attunements to the racial and class differences constitutive of premature and preventable death. Later in the essay, I elaborate the impact of COVID infection and mass deaths on sanitation and crematoria workers, often from "low" castes/classes, who bore the brunt of India's second wave. Elsewhere, I have pursued another wake in narrative and image for the uncounted migrant workers who trekked home after the draconian lockdown of March–May 2020.[10] My reflections here complement that exploration: the smoke signals from crematoria are as much about environmental pollution and mass death as they are about workers caught in smoke densities, vulnerable without personal protective equipment (PPE), open to manifold hazards, standing witness to the collective loss.

But it is not only these circumstances that relegate particular communities to premature and preventable death. Scholars like João Costa Vargas and Matthew Flynn, among others, analyzing death at mass scale during the COVID pandemic have shown how "social triage" in the pandemic emergency serves as an X-ray for normative organizational processes and forms affecting the distribution of life and death within so-called democracies.[11] Certain social congeries suffering from long histories of biopolitical neglect are already more vulnerable to the viral emergence than others; in the same vein, decolonial critics have tracked the necropolitical impacts of racial and class inequities that shape and define the biological existence of historically dispossessed communities. These conversations frame my claims of preventable and premature death, invoking a swathe of reports on differentials in COVID deaths that were not just about mismanagement—as in Trump's America, Jair Bolsonaro's Brazil, and Narendra Modi's India. By now there is a well-calibrated global picture of how the pandemic exacerbated normative necropolitics: those without long-term access to health care, nutrition, and housing died in greater numbers during COVID. In this regard, India's most vulnerable communities are no exception. More striking is the fact that, during India's second wave, no one from any class or caste stratum seemed to escape the clutches of virulence. This all-encompassing event fueled a ferocious outcry. After all, middle and upper classes and castes were dying without medical oxygen or access to hospital care: three of New Delhi's biggest private hospitals ran out of medical oxygen in April 2021 even as similar tragedies unfolded in Mumbai, Amritsar, Gurgaon, Kurnool, Moradad, Jammu, and Goa, to name a few local disasters of the early days.[12] Predictably, these impacts were exponentially higher for the most dispossessed.

Finally, my point of departure is India's second wave because the event foregrounded differential experiences within the otherwise globally synchronous event. Thinking differentially articulates necessary national specificities for a period when nation-state boundaries firmed up, governing resource distribution, health policy, and infrastructural management. Further, pandemic differentials attend to spatially localized effects emerging with disrupted connectivity—everything from transportation to global supply chains. While COVID was undoubtedly globally synchronous, the pandemic was experientially different in its intensities for formally or informally constituted local communities who had to make do with what they had where they were. In other words, even the historical particularity of national difference, resonant in "India's second wave," does not quite capture how the mass tragedy materialized differentially across the nation. The capital of the nation, Delhi, was one of the worst hit localities in part of because of the unavailability of direct pipeline supply to hospitals in this region of the country; most of the medical oxygen generating plants were located in the eastern parts. How, then, to conceptualize these cascading differentials? What critical analytic is both capacious and incisive enough for the task? I posit that explorations of COVID's "epidemic intensities" open to felt differences that, in turn, constitute the ground of respiratory politics.

Scholarly participation in the ongoing COVID wake recirculates images that qualitatively "account" for the uncounted dead as a matter of historical record; the accompaniment is visual analysis of what COVID image clusters convey in the media forms that carry them. To articulate what these forms "do" in relation to each other, I constellate a political aesthetic salient to Delhi's experience of COVID-19. Equally, I join creatives, activists, and scholars of all cloth invested in tracking epidemic intensities across the world.

The City and Bird: Communities of Air-Breath

One might begin with air as an extensive planetary environment that became visible in COVID's photojournalistic spreads, audiovisual documents, and climatic media. To begin with the latter, I draw on Yuriko Furuhata's *Climatic Media: Transpacific Experiments in Atmospheric Control* to offer a glance at COVID climatic media: those scientific and journalistic techniques and technologies that actively mediate and shape habitable environments.[13] As Furuhata notes, climatic media span a range that includes thermal-imaging photographs, charts, diagrams, and other such instruments of data visualization. Further still, pressing beyond conventional media objects like film, television, telephone, Internet and social media, Furuhata's media encompass natural elements like air or water. Sky media, for instance, are ensembles of atmospheric matter and their machinic inscriptions. During COVID, one classic genre of climatic media featured clean air images. These data visualizations attempted to explain what happened to the earth's atmosphere during the pandemic. Gathering information from two hundred continuous ambient air quality monitoring stations in 115 Indian cities, one coauthored paper published in July 2020 reported a 21–47 percent decrease in several elements such as nitrogen, carbon monoxide, and sulfur dioxide. The researchers transcribed city data to present national topoi: What happened to India's air, they wondered, after the March 25 lockdown? Crafted from satellite data recording the aerosol optical depth or the measure of aerosols (urban haze, smoke particles, desert dust, sea salt) distributed within a sample column of air, the data visualizations of India's air followed the conventions of the thermal-imaging photographs.[14] Encoding an environmentally auspicious atmosphere, a bright green patch spreading across India, these visualizations plotted clearing air as the lockdown slowed industrial production and fossil fuel consumption. As I have argued in *The Virus Touch*, scientific visualizations such as these are cultural performances in their mobilization of extant aesthetic grammars: in this case, green as the globally legible coding for better environmental futures.[15] Pitching the data for national green policies, the scientists offered the visualizations as imaginations of India's air futures. But even as these images speculated on emergent national surrounds, the planetary viewpoint from the satellite linked India's breathable air to universal air futures. Beyond tracking greener and greener air, the images further induced apprehensions of other times—that hoary past of "normal" times when industrial air sank into our lungs.

In such climatic media, we have an initial understanding of epidemic intensity. Inhabiting air-breath during COVID involved perceiving storied risk environments stretching from lung to body to room to county to globe. COVID infection made the fear of air rife with pathogens rushing into lungs an all-too-common experience. Even those who remained uninfected knew of this feeling from the sick and dying around them. Etymologically, intensity signifies an extreme stretching tight: it is a spatial notation of extensive infection topologies. As the air between us spreads but does not dissipate, intensity further suggests a piling on, an increase in volume: intensity bespeaks accumulation. Then the rush to render air-breath topologies calculable as risk environments: code red for dangerous air quality, green for breathable air. In the scientific images of climatic media, air-breath becomes calculable, acquiring technical valence. In this resonance, intensity is the measure for magnitude, degree, direction, or level of dilution. Air becomes comprehensible, laying ground for governability. But alongside this universalizing legibility, epidemic intensities index the deeply subjective character of acute infectious disease emergences. In the domain of feeling, intensity is the thickening/layering/bundling of sensations/affects; it is a term that translates qualitative perceptions of energetic forces between things into subjective experience. Breath is felt in the gasp of a lung: intensity is intimate. To speak of epidemic intensities, then, is to recognize layered, contiguous historical experiences constitutive of the global pandemic alongside the vast continuities of planetary infection topologies. Scientific, artistic, and popular images make epidemic intensities apprehensible, scattering breath into the air, infusing air into breath. Something latent, something imperceptible moves between us: we understand it, informationally; we sense it, affectively.

This brief excursion into the resonances of intensity frames my dive into the particularities of Delhi's air-breath experience. Since intensity is, finally, a comparative measure or feeling, a recognition of a rise and fall in states, epidemic intensities account for the differential vulnerabilities of the pandemic experience that are front and center in this volume on environmental justice. In terms of planetary topologies of air, the question of difference resides in scholarship on the varying toxicities of the chemosphere. Decolonial theorist Michelle Murphy's notion of the *alterlife* directs us to probe the parameters of molecular colonialism. Even as molecular scales decenter any illusion of mastery over the environment, argues Murphy, colonial, racial, military, gendered, chemical, biological, and geological structures constitute molecular life. The drift of PCBs, hormones, and soils may be often analyzed at molecular scale, but these assemblies are made of structural relations: some bodies bear more PCBs depending on in which neighborhood you live. Right before the pandemic, the Environmental Committee of the Forum of the International Respiratory Societies made the case for air pollution as one form of health inequality and environmental injustice arising from long-term inadequate health-care provision, food insecurity, unregulated housing, and toxic water infrastructure, to name just a few factors.[16] These structural forces, we might argue following Murphy, tune in to extensive and entangled relations: "Studying alterlife requires bursting open the categories of organism, individual, and body to acknowledge a shared, entangling, and extensive condition of being."[17] Even as we speculate our chemical or microbial relations, alterlife reorients us toward the possibility of "life otherwise," of another kind of future. One might wonder what that might be for India's COVID experience: What respiratory politics are afoot in imagining breathable air?

Climatic media suggests all creaturely life on the blue planet is at risk but to varying degrees. One genre of COVID lockdown feature photography made the case by folding representations of planetary media (earth and sky) into unusual creaturely activity. Wild or domestic livestock wandering into once-bustling environs induced laughter, wonder, sometimes shock. A pudgy sealion looked in askance at a shuttered shop; mountain goats marched down deserted streets; jackals howled in urban parks: the list mounted across media spreads. Such images espoused a familiar environmental politics: animals and birds are sentinels; their fate foreshadows ours; their re-inhabitation of planetary environs can anticipate other futures for all living species. These images were about lockdown effects: stilled industrial urban life, rewilded empty highways, deserted neighborhoods, and abandoned thoroughfares. But reminding humans of their creaturely existence, these feature photographs were also warnings: Will we, too, disappear, as "they" have in the age of mass extinctions as industrial metabolism eats into life-sustaining environments? Beyond this obvious apprehension, this image cluster constellated various "communities of air," locating them cartographically with captions and iconic signage. One spread in *The Guardian* gallery includes a photograph of peahens strutting about the Motilal Nehru Marg in New Delhi.[18] Consorts of the official "national bird" of India, these birds are memorable animal media in the archives of environmental politics. In one of Rahul Mukherjee's stories on environmental controversies concerning radiation of all kinds, he recounts one distrustful citizen of the western India city of Jaipur who believed the electromagnetic radiation from cellphone towers caused his brother's cancer. Even after protracted battles, he refused to believe scientific evidence of fading radiation. Rather, and rather poetically, it was the "return of peacocks to the garden" that restored his faith in emergent clean air.[19] Creaturely life, in this local story, once more appears as sentient guardians of the damaged planet. While I assemble these well-crafted photographs alongside scientific visualization, it should be obvious that the sensuous impact of these image clusters is significantly different. The melancholic, sometimes funny, photographs of creatures wilding urban or peri-urban milieus appeal to visual-haptic senses, orienting media consumers toward the planetary environs; by contrast, the scientific visualizations reach a narrower audience trained in reading abstractions and invested in the evidentiary. Whatever the differences, both render invisible "clean air" aesthetically perceptible—either by default (through creatures) or by color conventions.

Read together, the city and the bird come to form one part of Delhi's air-breath syntax; and this was the case well before COVID. For Delhi's urban pollution had been off the charts for decades. This air sensorium is the locus of Shaunak Sen's prize-winning documentary *All That Breathes*, filmed right before the pandemic. Seemingly, the documentary recounts the fortunes of two Muslim brothers who run a modest hospital in Delhi tending to injured birds, ill from industrial air. But other strands of audiovisual narration—the liveliness of creaturely life and the brooding prospect of Delhi's air—complement this human-centered story. As I have argued elsewhere, critical to the creaturely world is the amped up postproduction sound: the patter of feet, the squelch of larger bodies, squawks and snarls, delivering an off-screen world imperceptible to the eye, making creaturely existence viscerally palpable.[20] Visually, creaturely life—a snail, a turtle, a frog, centipedes and flies, monkeys, pigs—comes into view always in the foreground of rack focus shots. Meanwhile Delhi's human bustle, built structures, and sagging infrastructure loiter as blurry background.[21] The premiere protagonist of the creaturely world is not a friendly one: it is a ferocious kite swirling in the skies, adapting to but injured by Delhi's air, eating garbage and reducing the city's waste tonnage. One of the two brothers, Saud, speaks of the kite-human kinship as a "community of air" whose futures are inexorably bound, even as the visual syntax of iterative rack focus shots rolls humans, animals, birds, and insects into the air, water, and soil that sustains them. The more-than-human world is planetary but also inextricable from Delhi's historical situation of industrial calamity impacting humans, creatures, and the elements alike—perhaps at different velocities and intensities. Sen's respiratory politics is profoundly evident in the alterlife of Delhi's communities. In strange kinships, difficult and tenuous, we are reminded of what Murphy names "life otherwise."

Lung and Cylinder: COVID Breath Infrastructures

Climatic mediations of air as an extensive media environment are familiar to contemporary media studies.[22] But accompanying these mediations are *media intensions*, as Joshua Neves characterizes them, that index COVID-19's breath infrastructure, vital (the trachea, diaphragm, muscles, tonsils, lungs, mouth, and throat) and industrial (masks, hoses, oxygen tanks, medical equipment from the oximeter to the ventilator).[23] In this regard, pandemic images of the air-breath *complex* take us well beyond the air quality index. Inquiries into COVID breath hone the analytic gaze to local infrastructures and the differential vulnerabilities of the otherwise synchronic global airborne disaster.

The COVID pandemic has forced us to "see" air as an all-too-transitive medium for the transport of SARS-CoV-2.[24] Viruses lack locomotion; we know from epidemiological studies that viruses rely on air, water, or soil for transport, hitching a ride on life-sustaining environments for human and microbial life. This is why the air between us suddenly appeared as a dangerous medium as did the breath of the other—even at six-feet distances. Tied as they are to their site of original production, vital media like blood, feces, or breath were avowedly transitive, a realization reiterated during acute infectious disease epidemics; they habitually breach boundaries between individuals, between populations, and between species. The hyphen in the "air-breath complex" signals the contiguities of vital media, for breathing is not just embodied inhalation, as Tim Ingold notes, but also vaporized exhalation.[25] The idea of the "complex" further suggests the industrial infrastructure, those technological ensembles that regulate and condition air and breath; as psychological notation, the complex also implies the experiential dimension. The air-breath complex, then, indexes the continuities of a planetary medium (air) and the contiguities of a vital one (breath). We perceive a double movement, intensive (into the lung) and extensive (the air between bodies). One COVID image cluster highlighted this movement: these are the mask visualities that drew attention to the dynamic processes of breath becoming air and air becoming breath. During COVID, images of masks were global in their reach: one series located masking practices in particular places, especially cities with population densities, by featuring them in urban graffiti, billboards, and posters. Think Global Health's "Art of a Pandemic" feature, for example, devoted the visual spread to artistic commentaries on the state of masking all over the world;[26] in spreads such as these, the global synchrony of the COVID pandemic was on full display.

Moving inward, a second iconic image encoded what the mask sought to protect: the lungs, standing in for a vital infrastructure that includes the trachea, diaphragm, muscles, tonsils, mouth, and throat.[27] The lung erupted as the organ under threat in visualizations of physiological states in scientific and cultural notation. Some Indian cable channels promoting folksy remedies for COVID infection mounted breathing exercise videos, touting culturally Hindu solutions to SARS-CoV-2's onslaught on the lungs. One instructional demo featured a schematic of inhalation-exhalation: inhaling for measured seconds, then holding one's breath and exhaling in controlled counts. The video exemplified "media of breathing," as John Durham Peters characterizes them: in this case, the video promoted respiratory techniques culled from ancient yogic practices that train the body to transform air into breath.[28] When the video went viral on WhatsApp in 2020, national news channels quickly decried its contagious misinformation. If such lung therapy appeared possible in the early days of the pandemic, it was no longer tenable once India's medical oxygen crisis broke in 2021. And yet, in the midst of that later crisis, business-minded gurus such as Baba Ramdev persisted in touting yogic practices as ancient cures. Mocking those who gasped for air amid the unfolding tragedy of 2021, Ramdev pushed widely debunked plant-based cures for profit (more products in his Patanjali empire). "How can there be a shortage when God has filled the atmosphere with oxygen?" he argued. "Fools are looking for oxygen cylinders. Just breathe the free oxygen. Why are you complaining about shortage of oxygen and beds and crematoriums?"[29] Now it is the case that systemic Ayurvedic treatments are often egregiously sidelined as "complementary" to allopathic medical hegemony, but that was not the most chilling part of Ramdev's gambit. Rather it was his willed refutation of massive tragic losses that remains a spectacular instance of the necropolitical cynicism that had become the hallmark of the Modi regime. Ramdev's gambit underscored what some saw as fear and confusion, and others as callousness and indifference, from an authoritarian regime.

With the air thickening with crematoria smoke and bodies floating down the Ganges, another prosthetic arrived in Indian mediascapes: the medical oxygen cylinder. This part of the breath infrastructure made visible the intense breakdown of its vital counterpart—those beleaguered lungs no longer capable of transforming air into breath. Images pouring in from Delhi forged another syntax: the clunky industrial cylinder as the necessary complement to the failing lung. As images of heaving chests and gasping mouths proliferated on global media platforms, grassroots relief organizations, government organizations, and nongovernmental organizations hustled for medical oxygen. In these efforts, the medical oxygen cylinder instantiated a social rupture in the extreme: those who could afford it, commandeered it, while others slated for preventable and premature death became the countless "human interest stories" in the historical record of loss.

There are volumes written on the medical oxygen crisis and what it tells us about India's second-wave experience.[30] One might argue that the social rupture exacerbated by the peaking crisis of the second wave is an unsurprising scene in pandemic histories. Global public health emergencies habitually call the liberal-democratic bluff that values all human life equally. Pandemics emerge differentially for vulnerable communities who live with the structural violence of endemics (tuberculosis, malaria, dengue) and long-term social inequalities. And when resources are scant—as we saw with ventilators in the early days and later with vaccines distribution—a sorting and segregating of populations ensues. In this sense, India's second wave exemplifies the norm.[31] India was the second most affected country in world, accounting for one-seventh of the global COVID death burden. Yet this general paradigm cannot erase the absence of governance that exacerbated the crisis: Narendra Modi's government literally went "missing" for three weeks with government representatives protesting there was no insufficiency in the national medical oxygen supply. Close allies of the prime minister such as the thanatopolitical Uttar Pradesh chief minister Yogi Adityanath ordered property seizures of those who "spread rumors" about the lack of oxygen supply.[32] Amid cascading death, the Modi regime chose to spend their resources drumming up political victories and religious capital. Spectacular evidence of a callous realpolitik surfaced in the multiple super-spreader events of April 2021 such as huge election rallies to contest the West Bengal State elections and the massive Kumbh Mela gathering (that was expected to be convened in 2022). These actions or lack thereof turned India's second wave into a manmade disaster. While there was no accurate state data on the number of deaths during April–May 2021, a news report from the Institute for Health Metrics and Evaluation estimated the number of COVID-related deaths during April–May 2021 to be 650,000 (thrice the official number) and projected crossing one million by September 2021; other researchers modeling data culled from hospitals and nongovernmental agencies put the count at two million during this episode alone.[33]

In a naturalizing move, some blamed this disaster on the sudden emergence of an especially deadly and contagious mutation, the delta variant; moreover, they argued that it was the sheer scale and density of India's population that had turned raging infection deadly at mass scale. The government protested that the demand for medical oxygen had become unreasonable: from 3,842 metric tons (MT) a day at the beginning of April to 11,000 MT by the start of May. This was compounded by innumerable logistical challenges arising from modes of medical oxygen delivery to hospitals such as insufficient cryogenic tankers for liquid oxygen deliveries, inadequate filling stations for jumbo cylinders, and the safe transportation of medical oxygen tankers across state lines. A deeper read of the medical oxygen crisis, however, locates the problem in the Modi administration's selective approach to health expertise and their political commitment to private enterprise as the source of public goods. As early as September 2020, health experts had warned the administration that India's medical oxygen via direct pipeline was in short supply; if COVID infections were to increase, India was headed toward a catastrophe. Scientists marshaled data to produce dire estimates, forecasts, and predictions. With donations pouring into the PM CARES Fund after March 2020, in October, the administration floated tenders for the best bids for 162 pressure swing absorption (PSA) plants to manufacture pipeline medical oxygen.[34] Mired in financial bids, eight months later, none of these plants were as yet operational.

Under these circumstances, a brisk black market emerged around the medical oxygen cylinder as the most-vaunted industrial commodity. Across North, West, and Central India, small cylinders (that mountaineers use) were selling for 25,000 rupees ($300) in New Delhi while the standard cost was 1000 rupees ($12). Those who could afford it bought oxygen concentrators typically priced at 35,000–40,000 rupees ($420–80) for as much as 100,000 rupees ($1,200). The yawning social chasm broadened as the cylinder became iconic of the breath infrastructure, encoding industrial-modern solutions to a "natural" disaster. Concentrating "the basic part of breathable air," as Marjin Nieuwenhuis notes, the cylinder promised more than the lungs (which filter 21 percent of oxygen from the air) ever could; in this regard, the medical oxygen cylinder acquired the luster of a prized techno-fix.[35] But its value was equally economic because of scarcities and the problem of circulation. Despite the portability of the cylinder, supply chain disruptions and hoarding meant the cylinders were not easy to come by—a difficulty increasing exponentially for low-income wage earners. As newscasters pondered delays, detours, and supply chain logistics, the cylinder became *the* dominant icon of the breath infrastructure, infusing its image with desire.

Arguably, the cylinder is similar to Brian Larkin's "electricity box" as a media form that encodes "infrastructure"—at once a technological ensemble and a site of reciprocal exchange between humans and machines.[36] Ordinary infrastructure like electricity comes into view enformed; in this regard, the electricity box is a media form that constitutes a relation to otherwise invisible technological ensembles that condition industrial-modern existence. Most importantly, media forms, argues Larkin, are representational *and* experiential: they order matter even as their material configurations retain a concrete thingness. As such, we are aware of what is encoded in a media form and what is not "in" the frame as such: What technological ensembles constitute the cylinder we see in a photograph? What reciprocal exchange enables the material cylinder and its image to emerge? Further, Larkin's reading is generative because, like all infrastructure, the medical oxygen cylinder appears at the moment of failure. The iconic cylinder as valued commodity bespeaks both supply-chain failures and the malfunctioning lung.

Both invite a respiratory politics: a censure of absent governance that provided neither health care nor guaranteed remedy.

Across media streams, when the breath infrastructure was visible, it was at the brink of collapse, and spectacularly so, in lumbering oxygen tanks, in makeshift tubing, and in black fungus creeping along oxygen pipes. But in the midst of these images, one genre of photograph and video appeared consistently, intensifying semiotic resonances and affective investments in the cylinder image. This image cluster situated the medical oxygen cylinder in a variety of visual arrangements that all pointed up hoarding or scarcity. Photographs of cylinders stacked in storage rooms, rusty but ready, rubbed shoulders with their deployment in makeshift breathing outfits at Gurdwaras or in autorickshaws; a litany of media forms captured cylinders in transit on all modes of transport—in cars and taxis, clasped against bodies, rolled on carts. Like the electricity box, the cylinder drew attention to hitherto invisible technological ensembles and the logistics of delivery. Suddenly as necessary as lungs, the cylinder became the basis of a respiratory politics: Why weren't those PSA plants ready? Why did the government mismanage medical oxygen distribution? Why was medical oxygen not vital commons that, in Lauren Berlant's terms, holds and organizes "common life"?[37] Since the cylinder could not be disentangled from the lung, it induced sensory apprehensions of the very conditions of life—the *prana* and *pneuma* of ancient yore.[38] In all these ways, images of medical oxygen cylinders as COVID's industrial prostheses accumulated as intense affect. When it was visually missing in the countless pleas for medical oxygen circulating on WhatsApp, the cylinder signaled differential access to health care in a nation where less than 1–1.15 percent of the gross domestic product is consigned for the health sector.[39] Those reliant on state-run health care found themselves consigned to premature and preventable death.

With unerring accuracy, Arundhati Roy described the impossible demand placed on the most vulnerable at the mercy of the free market for their survival: "At the bottom end of the free market, a bribe to sneak a last look at your loved one, bagged and stacked in a hospital mortuary. A surcharge for a priest who agrees to say the final prayers. Online medical consultancies in which desperate families are fleeced by ruthless doctors. At the top end, you might need to sell your land and home and use up every last rupee for treatment at a private hospital. Just the deposit alone, before they even agree to admit you, could set your family back a couple of generations."⁴⁰ But who are the amorphous "they" slated for premature and preventable death? With the state refusal to count COVID-related deaths, what other indices signal "their" presence?

Smoke Signals: A Political Aesthetic of Air-Breath

One of the abiding mnemonics of India's second wave was the rising smoke from raging funeral pyres. In this genre of air-breath images, two strands dominate: aerial shots of crematoria from drone footage, representing the scale of mass death, and tableaux of hazmat-suited figures, barely visible, tending to pyres. Although culturally articular to Hindu death rites, these images encode loss. Like the bullet hole or weathervane gesturing toward a killing or a change in wind direction, in the famed Piercean instantiation of indexical signs, the smoke from crematoria invites a reconstruction of actual events; it gestures toward COVID-related deaths at mass scale. These smoke signals gain symbolic significance because of government refusals of a body count; in this accretion of meanings around a core graphic, the index becomes iconic.⁴¹

Many of the crematoria photographs are aesthetically impressionistic, the smoky thickness of their texture reminding us of those who died without

FIGURE 2. At the funeral pyre during India's COVID second wave. Anindito Mukherjee, May 9, 2021 (Getty Images).

oxygen.[42] Exquisite compositions of feature photography place spectral, PPE-clad figures at funeral pyres in close proximity to the vivid, licking flames bursting toward the photographic frame (see figure 2). The figures appear physically consumed by flames and hollowed by grief and loss as they perform the last rites; the photographs are often formally haptic, enclosing viewers "in" the intensities of heat. Arguably, the pandemic experience would intensify the sensory-affective force of the photograph: that is, the experiential intensities of mass death during the pandemic impact the viewing experience—then and now. This is even more the case for Indians within the country and in the diaspora, for few were left untouched by the tragedy. Quilting all others, partly because of their apocalyptic character, the crematoria images place us at the threshold of annihilation: at the extinction of air-breath. From the vantage of the pathological, we are forced to make new norms: What will those be? What future anterior or contingent present is still possible? Have we truly lost air/breath as social and ecological commons?

Impressionistic as they are, these photographs often incorporate spectral hazmat-clad figures watching over expiration. No doubt the humanoid figures serve as witnesses, anonymous filters for the photojournalist-turned-documen-tarian under conditions of informatic suppression. But they are also reminders of utterly hazardous labor, the smoke inhalation and risk of infection for those who work at the crematoria. With some crematoria recording thirty to forty pyres an hour at the height of the second wave, one can only imagine the extreme precarity of the labor despite the hazmat suits.[43] If the aerial images established the dangerous air quality for the neighborhood, then the images of witness situate viewers in the midst of smoky densities, inducing apprehensions of the gasping lung. In this regard, the crematoria images stitch together a visual syntax: the city, the sentinel (as human laborer), the lung, and industrial infrastructure (as crematoria). While cities with urban densities (the air sensorium), animals or birds as sentinels (communities of air), and (entropic) lungs as media of breathing are globally generalizable images of COVID's air-breath complex, the industrial infrastructure of life (medical oxygen cylinder) or death (crematoria) are locally particular notations of Delhi's second-wave epidemic intensities.

More importantly, further inquiry into who witnesses death tunes in to the question of environmental justice apprehended in this specific media form. For centuries, Dalit laborers have been assigned the labor of burning bodies. The smoke signals point to caste and class differences that define normally precarious circumstance—bodily injury from risk of infection and smoke inhalation—that become exponentially more dangerous at the peak of an airborne acute infectious disease emergence. The smoke signals connect the dots: they suggest varying epidemic intensities for those living within the same urban environs based on histories of dispossession. These varying epidemic intensities conceptually situate these images within ongoing critiques of racial capitalism.

Conclusion

In the absence of sustained data, inconceivable as that is for a regime whose proudest achievement is to digitize citizenship, there is sparse investigation into the implications of the epidemic for Dalit and Adivasi (Indigenous) communities who still remain at the lowest rungs of the socioeconomic ladder. Here, Black death as ontological condition returns as a generative analytic. To be sure, anti-Blackness speaks to the singularity of the Black experience under racial capitalism. Yet it is a viable starting point for examining the distributions of premature and preventable death during COVID because those arose from decisions to let die built on extant historical differences—racial, ethnic, religious, sexual, and otherwise. In the Indian case, the lockdown decision cast migrant workers out of the fold of state dispensations in what I have characterized as population culling.[44] Grassroots organizations catalogued the racial contours of who died on the long march home. In these killing fields, the labor of sanitation (as cleaners, manual scavengers, and waste pickers) and at the crematorium exposed Dalit and Adivasis to greater pandemic risks.[45] In the intensification of precarity, we see normative necropolitics become virulent. The analysts worry about health-care access in the rural pockets in India's six hundred thousand villages where Dalits live, not to mention caste discrimination that lessens chances of receiving care even where available; black-market prices further exacerbated the deathly consequences of health-care access paucities.[46] The crematoria work—that hazy figure as mute witness—compounds precarity, signaling the broadening racial divide. And so, the wake begins for the uncounted dead, for those at greatest risk of bodily injury from smoke inhalation slated for premature death.

The smoke signals, once a (preindustrial) part of the communication infrastructure, connect photographs emerging from Delhi's second-wave experience to greater COVID losses in historically dispossessed communities. In the modern era, one may recall *Smoke Signals* (1998) was one of the earliest independent (U.S.) Indigenous films based on the story of famous Native American writer Sherman Alexie. The icon accrues resonances even as it blurs, makes illegible, who and what is "in" the picture. Left to our imagination, we set about the cultural work of apprehension as the bases for a critical respiratory politics. That politics urges pragmatic action: everything from policies on air pollution to meeting medical oxygen manufacture thresholds to more investment in health care and streamlined free health-care access based on income differentials. Beyond practicalities, thornier concerns urge concerted social action to oppose and uproot caste and class discrimination that has only increased with the tightening grip of the Hindu Brahmanical minority at the helm of Modi's India. The arc toward environmental justice is in sight: it points toward the labor of the political aesthetic.

The visual syntax for India's second wave serves up a warning: city + bird + lung + cylinder = smoke signals. It should fill us with apprehension. At an eschatological threshold, smoke compels intuition: the quivering awareness that there is no longer a guarantee that the mediums that sustain us will survive for the damage we have wrought is catastrophic. We are out of breath, as indeed, out of air.

NOTES

1. Arundhati Roy, "The Pandemic as Portal," in *Azadi: Freedom, Fascism, Fiction* (London: Penguin Random House, 2020), 214.
2. Siddiqui won the 2018 Pulitzer Prize for Feature Photography covering the Rohingya Crisis. A few years later in 2021, he was killed while covering Afghan and Taliban force skirmishes on the border of Pakistan.
3. Arundhati Roy, "We Are Witnessing a Crime against Humanity," *The Guardian*, April 28, 2021, https://www.theguardian.com/news/2021/apr/28/crime-against-humanity-arundhati-roy-india-covid-catastrophe.
4. Simon Scarr et al., "COVID-19 India: Mass Cremations, Day and Night," Reuters Graphics, May 28, 2021, https://www.reuters.com/graphics/HEALTH-INDIA/CORONAVIRUS-DEATHS/qzjvqrqaqpx/.
5. Zahid Chaudhry, *Afterimage of Empire: Photography in the Nineteenth Century* (Minneapolis: University of Minnesota Press, 2012).
6. In *On Populist Reason* (London: Verso, 2005), Ernesto Laclau outlines the operations of

populism through affective investments in symbols (like "Joe the Plumber," for instance) that organize heterogeneous social demands.

7. Brian Larkin, "Promising Forms: The Political Aesthetics of Infrastructure," in *The Promise of Infrastructure*, ed. Nikhil Anand, Akhil Gupta, and Hannah Appel (Durham, NC: Duke University Press, 2018), 327–43.

8. Jacques Rancière famously develops "the partition of the senses" in *The Politics of Aesthetics*, trans. Gabriel Rockhill (London: Bloomsbury Publishing, 2006).

9. Christina Sharpe, *In the Wake: Blackness and Being* (Durham, NC: Duke University Press, 2016), 12.

10. See Bishnupriya Ghosh, "The Plague Check: Population Culling as Pandemic Realpolitik," *Catalyst: Feminism, Theory, Technoscience* 10, no. 1 (2024): 1–23.

11. See João Costa Vargas, "Blue Pill, Red Pill: The Incommensurable Worlds of Racism and Anti-Blackness" (183–205) and Matthew Flynn, "Global Capitalism, Racism, and Social Triage during COVID-19" (206–20), essays published in the special issue of *Kalfou* 8, no. 1/2 (2021). In identifying COVID measures as social triage, Flynn offers a range of pandemic situations where premature and preventable death became part of the calculus: his instances range from meatpacking industry workers in the United States to Asian migrant workers in Saudi Arabia's *kafala* system, all experiencing high and rising infection rates because of working and living conditions.

12. See Aditya Bahil, "Breathless India," *Sidecar*, May 27, 2021, https://newleftreview.org/sidecar/posts/breathless-india.

13. Yuriko Furuhata, *Climatic Media: Transpacific Experiments in Atmospheric Control* (Durham, NC: Duke University Press, 2022).

14. See Ram Lal Verma (Asia-Pacific Network for Global Change Research) and Jatinder Singh Kamyotra (director of the Central Pollution Board in Delhi), "Impacts of COVID-19 on Air Quality in India," *Aerosol and Air Quality Research* 21, no. 4 (April 2021): 1–19.

15. Bishnupriya Ghosh, *The Virus Touch: Theorizing Epidemic Media* (Durham, NC: Duke University Press, 2023); see chap. 2's discussion of aesthetic grammar and scientific visualization.

16. Dean E. Schaufragel et al., "Air Pollution and Non-Communicable Diseases: A Review by the Forum of International Respiratory Societies Environmental Committee Part I: The Damaging Effects of Air Pollution," *CHEST* 155, no. 2 (2019): 409–16.

17. Michelle Murphy, "Alterlife and Decolonial Chemical Relations," *Cultural Anthropology* 32.4 (2017): 494–503, https://journal.culanth.org/index.php/ca/article/view/ca32.4.02.

18. "The Urban Wild: Animals Take to the Streets amid Lockdown," *The Guardian*, World Gallery, April 22, 2020, https://www.theguardian.com/world/gallery/2020/apr/22/

animals-roaming-streets-coronavirus-lockdown-photos.

19. Rahul Mukherjee, *Radiant Infrastructures: Media, Environment, and Cultures of Uncertainty* (Durham, NC: Duke University Press, 2020), 24.

20. Bishnupriya Ghosh, "All That Breathes," *Docalogue*, https://docalogue.com/all-that-breathes/.

21. Ghosh, "All That Breathes."

22. Contemporary expansions of the media concept articulate media in double senses: as life-sustaining mediums (air and water, blood and breath) and as technical mediums that render human and nonhuman signals readable. There is robust scholarship on the many conjunctures of media and environment: on the deleterious dimensions of media in the environment (sonic booms to toxic e-waste); on how media scale between local events (a flood, a storm) and planetary processes (sea-level rise); on how media record and track ecological relations and processes, readying them as environments in need of intervention; and what media can do to slow down, redirect, even thwart ongoing planetary damage. In all these accounts, mediatic processes compose what we understand as the "environment."

23. Joshua Neves questions the insistent "*externalization thesis*," an onto-epistemological framing perhaps most commonly associated with Marshall McLuhan's 'extensions of man.'" He notes: "While this material and imaginary vector is no doubt important (and more nuanced than I can develop here), it also drives a set of assumptions that inhibit alternative ways of knowing media. My argument here is this: it's time we give more attention to the *intensions of (hu)man*. I have in mind processes of coming into the body, including pharmacological habits like taking a pill or applying electrical current. But also, a broader set of concerns related to eating, ingestion, absorption, inhalation, injection, and so on." "Technology + Pharmacology: Notes on Current Research, *Heliotrope*, September 15, 2021, https://www.heliotropejournal.net/helio/technology-pharmacology. See also Joshua Neves, "The Internet of People and Things," in Neves et al., *Technopharmacology* (Minneapolis: University of Minnesota Press, 2022), 89–119.

24. This true of all airborne epidemics hailing back to Hippocrates who understood air to be the conduit of life and of disease: see discussion in Catherine Albano, *Out of Breath: Vulnerability of Air in Contemporary Art* (Minneapolis: University of Minnesota Press, 2022), 5.

25. Tim Ingold, "On Breath and Breathing: A Concluding Comment," *Body & Society* 26, no. 2 (June 2020): 158–67.

26. Mary Brophy Marcus, "The Art of a Pandemic," Think Global Health, November 5, 2021, https://www.thinkglobalhealth.org/article/art-pandemic.

27. The lungs were established as part of breathing's vital infrastructure with the pneumatic

experiments of the late sixteenth century. See Jean-William Fitting, "From Breathing to Respiration," *Respiration* 89, no. 1 (2015): 82–87.

28. John Durham Peters, "Media of Breathing," in *Atmospheres of Breathing*, ed. Lenart Škof and Petri Berndtson (Albany: SUNY Press, 2018), 179–98.

29. "Ramdev: Doctors Furious over Yoga Guru's Insulting Covid Remark," *BBC News*, May 25, 2021, https://www.bbc.com/news/world-asia-india-57237059.

30. The spread is heterogeneous, ranging from news items during the crisis to special reports by research institutions in the aftermath. See, for instance, Simran Sirur, "How Medical Oxygen Is Supplied to Hospitals, and Why India Is Facing a Shortage," *ThePrint*, April 27, 2021; Moonis Mirza et al., "India's Multi-Sectoral Response to the Oxygen Surge Demand during the COVID-19 Pandemic: A Scoping Review," *Indian Journal of Community Medicine* 48, no. 1 (January–February 2023): 31–40; and Amit Thadani, "Preventing a Repeat of the COVID-19 Second Wave Oxygen Crisis," *Observer Research Foundation Report* 151 (2021): 1–17, for a selection of coverage.

31. In January 2023, the annual session of the World Health Organization (WHO) adopted a resolution for affordable provision of medical oxygen across the Global South: "Increasing Access to Medical Oxygen," WHO Executive Board, January 31, 2023, https://apps.who.int/gb/ebwha/pdf_files/EB152/B152_CONF4-en.pdf.

32. See Omar Rashid, "Oxygen Shortage," *The Hindu*, April 25, 2021, https://www.thehindu.com/news/national/other-states/seize-property-of-those-spreading-rumours-up-cm/article34404518.ece.

33. See Subramanian Natarajan and Poonam Subramanian's "Systematic Review of Excess Mortality in India during the COVID-19 Pandemic," *Indian Journal of Community Medicine* 47, no. 4 (October–December 2022): 491–94, for a complex account of excess mortality (compared to normative mortality rates) and estimates of COVID deaths up until June 30, 2021, despite the egregious underreporting.

34. These plants absorb nitrogen from ambient air to concentrate oxygen for pipeline delivery to hospitals.

35. Marjin Nieuwenhuis argues that oxygen was the modern answer to the debates over the "pnuema" as the interface of air and the respiratory system. Named in 1789, the discovery of oxygen made for a biopolitics of air as it became possible to govern breathability. In the use of poison gas, knowledge of air would become the basis of a thanatopolitics. See "The Politics of Breathing: Knowledge of Air and Respiration," in Škof and Berndtson *Atmospheres of Breathing*, 199.

36. See Larkin, "Promising Forms."

37. Lauren Berlant, "The Commons: Infrastructures for Troubling Times," *Environment and*

Planning D: Society and Space 34, no. 3 (June 2016): 393–419.

38. See Albano, *Out of Breath*, 6–7.

39. Ruby Dhar et al., "Faultlines in India's COVID-19 Management: Lessons Learned and Future Recommendations," *Risk Management and Healthcare Policy* 14 (2021): 4382.

40. Roy, "We Are Witnessing a Crime against Humanity."

41. See Bishnupriya Ghosh, *Global Icons: Apertures to the Popular* (Durham, NC: Duke University Press, 2011), chap. 2 on icon semiosis.

42. See "India Coronavirus: Round-the-Clock Mass Cremations," *BBC News*, April 28, 2021, https://www.bbc.com/news/in-pictures-56913348.

43. WHO's models of burning pyres noted that, during the second wave of COVID-19 in 2021, surface ozone levels were observed to be higher, which had a direct impact on lung function, thereby making people more susceptible to COVID-19. See M. G. Manoj et al., "Exacerbation of Fatality Rates Induced by Poor Air Quality Due to Open-Air Mass Funeral Pyre Cremation during the Second Wave of COVID-19," *Toxics* 10, no. 6 (2022): 1–10.

44. See Bishnupriya Ghosh, "The Plague Check" and "The Refugee within: Becoming Disposable in Modern India," *The Large Glass: Journal of Contemporary Art, Culture and Theory* 29/30 (2020): 104–14.

45. Of India's 1.4 billion, 25 percent of the population are Dalit and Adivasi: Priyali Sur, "Under India's Caste System, Dalits Are Considered Untouchable: The Coronavirus Is Intensifying that Slur," CNN, April 16, 2020, https://edition.cnn.com/2020/04/15/asia/india-coronavirus-lower-castes-hnk-intl/index.html. Further, according to the Socio-Economic and Caste Census 2011, 182,505 households in rural areas reported themselves as manual scavengers or sanitation workers, and this figure would be elevated substantially if we identified the households engaged in manual scavenging in urban areas. Of the six million households, 40–60 percent are made of Dalit subcastes: "Manual Scavenging," Press Information Bureau, Ministry of Social Justice & Empowerment, Government of India, December 8, 2015, https://pib.gov.in/newsite/printrelease.aspx?relid=132739.

46. Priyali Sur's "Under India's Caste System" quotes Paul Divakar, a Dalit activist from the National Campaign on Dalit Human Rights: "India has 600,000 villages and almost every village a small pocket on the outskirts is meant for Dalits. . . . This settlement is far from health care centers, banks, schools and other essential services. During times like Covid-19, the aid may not even reach this small pocket."

Fueling Toxicity

Fast Fashion, Air Pollution, and Slow Violence

Savannah Schaufler

As I write this essay, I am reminded of my experience in Nicaragua, where I watched the delivery of large bulks of clothing, presumably from Western charity bins. This experience allowed me to understand that clothing is not just a commodity or an expression of identity. It carries narratives of superiority and colonial power, intertwined with the global phenomenon of fast fashion. And it opened my eyes to the hidden power dynamics at play, where some regions bear the burden of waste and pollution while others enjoy the consumption of cheap fashion.

The global fashion industry, with its relentless demand for rapid production and consumption, has led to a staggering ninety-two million tons of clothing ending up in landfills worldwide, leaving behind a trail of negative social and environmental consequences.[1] This essay uses the Kantamanto Market in Accra, Ghana, as an example to explore the realities of textile toxicity and the far-reaching implications of the fast-fashion industry. Regardless of the location of their original entry to the global market, many textiles eventually find their way to this bustling secondhand market. Here, masses of clothes from clothing bins and charity collections mainly from the United States and countries in Europe

47

are sent for resale and reuse. While the United States and Europe aim to recycle their fast-fashion textiles, Ghana bears the brunt of various aspects of pollution. Mountains of worthless and mainly inferior-quality clothing, consisting of plastic fibers that cannot be reused and resold, are incinerated at dumpsites near the Kantamanto Market, polluting the air with noxious fumes.[2] The incineration of clothing is a major contributor to air pollution, especially due to the synthetic chemicals and dye toxins used in the manufacture of clothing.[3] As the fashion industry thrives on rapid production and consumption, it poses significant eco-logical threats to human and planetary health. However, to understand the roots of this ecological crisis, it is necessary to venture back to the Industrial Revolution, which transformed clothing production and consumption, establishing the textile industry as a cornerstone of the economy.[4] This transformative period, marked by the invention of the sewing machine and other technological advances, including the development of nylon, the first fully synthetic fiber, by American chemist Wallace Carothers in 1935, and the creation of polyester by British chemists John Rex Whinfield and James Tennant Dickson in 1941, reshaped the landscape of clothing production and consumption.[5] Polyester, which gained popularity in the 1970s, was established as a symbol of progress, contributing to the era's shiny suits and establishing itself as a cheap fabric.[6] Prior to these innovations, clothing and textiles were crafted through artisanal processes, albeit with significant labor intensity, making new clothing a privilege of the upper classes.[7] At the same time, this technological evolution revolutionized labor patterns and prompted im-provements in transportation and communication systems, thereby democratizing the availability of clothing.[8] By the mid-nineteenth century, fashionable clothing was affordable to citizens of various backgrounds, sparking an unprecedented boom in clothing production and eventually replacing durable garments with disposable ones.[9] This development marked a significant turning point in the history of fashion and its environmental consequences.

This essay explores the fast-fashion industry's disruptive impact on social, economic, and ecological factors, particularly its contribution to air pollution. The term "fueling toxicity" highlights how the fashion industry's processes and practices exacerbate air pollution, posing threats to the environment and human health. It will unfold in roughly two parts: First, attention will be drawn to the theoretical framework in which this exploration is embedded. This involves introducing the "actor-network theory" (ANT), a conceptual framework crucial for understanding the web of relationships and dependencies involved in the globalized production, consumption, and disposal of clothing. ANT functions not merely as a framework but as a guiding philosophy for interdisciplinary methods. This approach, derived from science and technology studies, helps to decipher the multiple relationships at play within the fast-fashion system. ANT, as both a conceptual and methodological tool, serves two essential purposes: first, it extends the analysis beyond individual actions to encompass the broader network of people and elements that collectively shape the fast-fashion system, and second, it uncovers previously unnoticed actors, networks, and historical patterns involved in the secondhand clothing business that often remain overlooked in the Global North. The application of ANT allows us to move beyond simplistic understandings of waste as a consequence of human behavior, recognizing how various actors contribute to the fast-fashion industry's toxic waste problem. Drawing on ANT as well as on the concept of "waste regimes" put forward by sociologist Zsuzsa Gille, the essay examines the production and consumption of fast fashion as an interconnected network of actors and explores the globalized waste of clothing and its impact on the environment.[10] "Waste regimes" address the materiality of waste that is embedded in and shaped by social relations, power asymmetries, and economic processes.[11] The second part engages with toxic embodiments through examining how exposure to toxins intertwines with time, space, and the body, particularly in relation to the colonial legacies and power structures that shape the fast-fashion industry.[12] Therefore, it explores how these temporal worlds and scapes manifest in the realities of aerial health disparities among diverse populations, emphasizing the concept of "slow violence" proposed by environmental humanities scholar Rob Nixon as a lens to identify the hidden power dynamics at play.[13]

Using these methodologies, I aim to explore the intricacies of fast fashion, its connection to air pollution, and the structures of harm and inequality it perpetuates. By examining the realities of different populations, I intend to contribute to a comprehensive understanding of the environmental, social, and ethical dimensions entangled within the toxic threads of secondhand clothing. While secondhand clothing is often seen as a resource for low-income countries, it is important to recognize that it can also pose significant environmental and health challenges, effectively becoming a form of toxic waste. In terms of the language used in this essay, it is important to acknowledge that the notion of "we" has long been attributed to the entirety of humanity, a collective responsible for the state of the planet.[14] However, embedded within this seemingly universal narrative are layered power dynamics and disparities that often remain unaddressed. Conventional discourse on the climate crisis, through its use of the terms "we" and "us," assumes an equal distribution of responsibility and overlooks the nuanced influences of power structures.[15] This essay acknowledges the homogeneity of this "we," using it not as a sweeping generalization but as a means to engage with readers. In this context, the "we" becomes a reference to Western-oriented, industrialized nations and stakeholders enmeshed in practices like outsourcing and the secondhand textile industry. This perspective illuminates the fact that culpability for the environmental crisis is not universally shared; rather, it is concentrated within specific spheres defined by historical legacies and economic systems.

Yet, before discussing the secondhand clothing business and its role in shaping multiple realities of aerial health, it is essential to acknowledge my position and perspective within this context. While the Kantamanto Market in Ghana serves as an example to discuss the complexities of the secondhand clothing trade, I have not personally visited the market. The information presented in this essay is solely derived from online sources, academic references, and scholarly works. My analysis of the issues surrounding the fast-fashion industry and the globalized secondhand clothing trade, as exemplified by the Kantamanto Market, is not limited to Ghana alone. The challenges of constant overproduction, waste colonialism, and environmental injustices extend beyond this particular market, reaching other countries as well. I recognize that my perspective as a researcher is shaped by my background and positionality. As a white female early-career researcher from an upper middle-class Austrian society, I aim to approach this topic with sensitivity and awareness of the historical and structural power dynamics at play in the globalized fashion industry. By acknowledging my point of view, which includes recognizing that I am myself also part of the problem as a consumer contributing to the fast-fashion industry, I want to facilitate a broader conversation about the need for sustainable and equitable practices in the global fashion industry, and to contribute to a better understanding of the environmental and social complexities inherent in the fast-fashion and secondhand clothing trade. Despite this, it is important to emphasize the diverse and multiple lived realities of individuals who deal with the consequences of excessive clothing waste and work in and around the market.[16] In the following sections, the convergence of various nodes of waste and its acting entities will be explored, guided by a theoretical framework that addresses the ramifications of the toxicity and materiality of used clothing, with a particular focus on its connection to air pollution.

Theoretical Framework: Nodes of Waste Relationality

The secondhand clothing market transcends mere resource utilization for certain actors; it also encapsulates a complex interplay of violence and toxic waste. Exploring some conceptual frameworks used by scholars working with ANT, along with critiques and concepts such as sociologist Zsuzsa Gille's "waste regimes," allows us to delve deeper into the multifaceted (socio)economic, structural, ecological, and colonial aspects of the secondhand market. Recognizing the involvement of numerous actors in this process, a critical examination of everyday life, particularly within Western and industrialized nations, gains significance.[17] This is crucial as the interconnectedness of the natural world and humanity is undeniable, resulting in reciprocal influences, both positive and negative. However, within the realm of political ecology, this interconnectedness also unveils pressing issues, such as extensive waste generation, social inequality, and injustices.[18] Acknowledging the ubiquity of waste in everyday life, regardless of location or social constructs, is essential to understanding waste as a concept in itself. Waste serves as a constant companion. Nevertheless, as environmental engineering scholar Eva Pongrácz notes, "Waste is a value concept, culturally constructed and subjective to the individual, be it the observer or the deposer." The value ascribed to waste fluctuates depending on one's circumstances, geographic position, cultural identity, and interpretations. Thus, waste, as it relates to humanity, defies objectivity, intertwining closely with subjectivities.[19] Sociologist Martin O'Brien further examines the value of waste within social contexts and the power structures it encompasses.[20] According to him, waste is not solely a product of disregard; it is also subject to a sophisticated interplay within the web of public and private networks. O'Brien's insights shed light on the multifaceted nature of waste, where what one group discards as excess can hold value for another.[21] This assertion, however, requires critical examination through the lens of colonial structures and power imbalances. In the context of the secondhand industry, the saying "one man's meat is another man's poison" finds resonance.[22] Waste from more affluent societies, discarded like excess, becomes a resource for those in less privileged circumstances. This juxtaposition reveals the power dynamics that Western nations maintain over less industrialized countries. This power enables them to offload their waste, thereby avoiding the responsibility of managing it within their own borders. Unfortunately, this process not only impacts the natural environment; it imposes a heavy toll on the inhabitants of these recipient nations. The influx of waste engenders a cycle of exploitation and environmental degradation, where what is discarded as unwanted in one place becomes a source of harm and injustice in another.

Humanities scholars such as Massimo De Angelis, Marco Armiero, Zsuzsa Gille, Max Liboiron, and Gay Hawkins have explored the complex intersections of waste, pollution, and colonialism. Their contributions, like De Angelis and Armiero's conceptualization of the "Wasteocene," which navigates the socioecological dynamics of power and toxicity, can give rise to narratives about marginalized communities and neglected environments.[23] Gille's concept of "waste regimes" considers an intertwined, global relationship of waste through the lens of social institutions and conventions, its perception and politicization.[24] Liboiron's incisive study of pollution's colonial dimensions offers a de- and anticolonial perspective, reshaping the conventional understanding of pollution.[25] And moreover, Hawkins's exploration of disposability provides insight into a fundamental element of modern capitalist systems.[26] Through their scholarly contributions, these researchers have brought to light the complex historical, social, and environmental dimensions that weave waste, pollution, and colonial power into a nodal network. However, in the context of the networks and diverse actors involved in the fast-fashion industry, especially in the redistribution of textile donations or "waste," a more precise theoretical framework becomes essential.

ANT offers here an interdisciplinary perspective, which envisions the world as a complex network interweaving humans, objects, ideas, and concepts. This theory embraces and contemplates the influence of more-than-human actors within this complex framework.[27] The main point of ANT lies in the convergence of different elements, culminating in more or less coherent actors, that constitute its main study.[28] ANT perceives agency as a distributed performance that arises from the interplay between various entities, encompassing both human and more-than-human.[29] Exploring the practices that emerge in diverse locations, spaces, and over time, ANT identifies the entangled dynamics of human and more-than-human entities that enable (inter)actions.[30] This theory proves particularly useful in exploring the impact of networks on the realm of space, bridging the gap between nature and society, and revealing how materials significantly shape the world.[31] As outlined by sociologist and psychologist Bruno Latour, ANT rejects traditional dimensional thinking and adopts a "nodes" perspective, according to which each entity has an equal number of dimensions and connections, negating the need for proximity/distance as the sole determinant.[32] Latour argues that one network is not larger or smaller than the other; instead, each network varies in length and intensity of connections.[33] Within ANT, it is essential to redefine the notion of actors, as it extends beyond purely human agents. Here, an actor represents a semiotic definition, signifying something with the capacity for action or agency attributed by others.[34] Rather than being the instigator of action, an actor-network serves as the dynamic destination of a vast array of entities.[35] ANT's claims go beyond macro and micro distinctions, while scholars, such as Zsuzsa Gille, critique this view and argue for the analytical independence of these levels.[36] In her work, Gille seeks to merge the Marxist concept of the mode of production with ANT's concepts, collectives, and sociomaterial formations to construct a macrotheoretical framework for studying the interplay between waste and society.[37] While maintaining the separation of macro and micro perspectives, Gille retains the emphasis on material concreteness in ANT for the specific purpose of theorizing waste. For Gille, waste exists within a cultural and hybrid context, encompassing both human and more-than-human elements.[38] Consequently, Gille situates her concept "waste regimes" within the context of waste, asserting that the solution lies in shifting levels of abstraction rather than altering analytical perspectives.[39]

In her explorations, Gille operates at the macro level, analyzing the production, circulation, and transformation of waste as a concrete material. This approach is concerned with the social institutions, referred to as regimes, that determine the perceived value of different types of waste or resources. Regimes also govern the production and distribution of waste. By considering social relationships and material interconnections, the concept of "waste regimes" addresses questions related to waste production, representation, and politics. This approach fosters an understanding of the economic, social, and cultural underpinnings of particular forms of waste, as well as the underlying rationale for their production.[40] Such interconnections resonate with the dynamics observed in the secondhand clothing industry, where economic, social, and cultural structures intertwine with waste generation, politics, production, and distribution. Thus, the secondhand textile industry finds a suitable theoretical framework within the concepts discussed in this section, allowing for a careful analysis intertwined with sociopolitical considerations.

Toxic Narratives and Colonial Dynamics of Clothing Waste

Consumers in the Global North are encouraged to transcend the mere act of discarding their old clothes and donate them, aiming to avert resource depletion. Aligned with charitable organizations, the collection systems eagerly advocate for donors to generously contribute their unwanted and old clothing, thereby providing support to the "underprivileged and needy." However, this display of charity is often limited by the constraints of convenience.[41] This discourse on textile recycling practices has a strong resonance in northern Europe and is influencing the emergence of public policies that aim for an environmentally and socially sustainable use of textile resources and waste.[42] Nonetheless, while recycling and donating are praised as an ethical behavior to inspire benevolence, its impact on the subsequent market exchange in this second life remains rather marginal.[43] Remarkably, a disturbing "out of sight, out of mind" attitude prevails among various actors, who seem to ignore the far-reaching implications, geographical distribution, and socioeconomic impact of trade dynamics in less economically advanced countries.[44] Ghana, alongside other countries, including India, Malaysia, Pakistan, Kenya, and Cambodia, plays a pivotal role in the global import of secondhand clothing intended for resale and reuse, and the Kantamanto Market in Accra stands as the central hub.[45] Flourishing within the city's informal economy, this market specializes in the trade of used clothing primarily sourced from Western countries. An astonishing 15 million kilograms of clothes exchange hands on a weekly basis, culminating in a staggering annual total of approximately 150 million kilograms.[46] The operations at Kantamanto, employing around thirty thousand individuals, involve various tasks, ranging from cleaning and repairing to recycling and selling the imported garments.[47] Nevertheless, the fact that a significant portion, estimated at 40 percent, eventually finds its resting place in landfills is a poignant reminder of the devastating waste generated by the fast-fashion industry and its attendant negative environmental consequences.[48]

At the same time, there seems to be a parallel decline in African textile production.[49] Thus, the Kantamanto Market not only serves as a commercial outlet, but also embodies the pressure to align with contemporary clothing aesthetics, frequently dismissing traditional clothing as "inappropriate."[50] In this context, the Ghanaian term "obroni wawu," signifying "dead white man's clothes," takes center stage.[51] The term alludes to the perception that the clothes found at the Kantamanto Market, once owned by (deceased) white individuals, carry connotations rooted in colonial legacies and the asymmetrical power dynamics propagated by white supremacy.[52] This translation highlights the disconnect between Ghanaians and the garments they acquire, underscoring that these clothes were never intended for them to wear and were never tailored to their needs. Rather, they are perceived as commodities accessible primarily to white individuals.[53] Yet the seemingly altruistic practice of donating clothes in Western countries demands critical scrutiny through a colonial lens, particularly within the domain of the secondhand clothing market. This market often perpetuates a toxic narrative of allegedly helping the "poor and needy" individuals in Africa with hand-me-downs, obscuring the reality that countries in the Global South bear the burden of clothing waste due to structurally oppressive systems.[54] While the consumption of low-quality fast fashion surges in the Global North, the repurposing of used clothing as commodities for the Global South perpetuates ongoing dependency structures between these spheres, fostering persisting inequality in these nations.[55]

In these nodes of relationships explored within ANT, the concept of "toxic timescapes" plays a crucial role in illustrating the unequal distribution of environmental burdens across different communities and regions.[56] Within the ANT framework, "toxic timescapes" emphasize the temporal dimensions of the environmental impacts generated by human and more-than-human actors. Environmental injustices frequently subject specific populations to disproportionate exposure to toxic substances and hazards, thus exacerbating the repercussions of "slow violence" on marginalized groups.[57] This paradigm focuses on the interplay of time, space, and bodies in relation to toxic exposure. Rooted in environmental humanities, "toxic timescapes" portray the insidious impact of environmental hazards and toxic substances on both humans and more-than-humans over extended periods. The complex interconnections between toxicity and pollution, visible through the prism of power dynamics on individual and collective levels, pervade human and more-than-human spheres. "Toxic timescapes" emphasize that exposure to toxins and pollutants is not merely a spatial concern but is deeply entwined in time. "Toxic timescapes" highlight the temporal dimensions of environmental degradation and how the effects of toxic substances unfold over time, leaving a lasting imprint not only on concrete spaces, but also on landscapes and communities more broadly. A relationship emerges between these timescapes and the environmental challenges arising from the advances of industrial civilization, exemplified by the fast-fashion phenomenon and the material flow of textile waste.[58] Here, the concept of "slow violence" posited by Nixon employs the gradual and often imperceptible violence perpetuated by environmental destruction and toxic contamination across temporal and spatial dimensions.[59] "Toxic timescapes" cast a light on the disparate distribution of environmental burdens across communities and regions, whereby environmental injustices result in specific populations bearing an unequal burden of toxic substances and hazards, intensifying the impact of "slow violence" on marginalized groups and bodies.[60] In alignment with these perspectives, environmental and political geographer Thom Davies engages with the notion of "slow observation," an endeavor borne from years of immersion into a place, unfolding through time.[61] This concept elucidates how communities dealing with pollution internalize the lived reality of ongoing environmental hazards.[62] Davies interweaves the idea of "toxic timescapes" with ethnographic observations within an industrialized region in Louisiana, casting light on the precarious intersectionality of time, space, and corporeal existence amid a landscape of industrial pollution.[63] "Toxic timescapes," therefore, recognize that environmental harm and the impact of toxic elements on airscapes and landscapes, as well as human and more-than-human health, are not static or confined to a specific moment but manifest themselves as dynamic processes over time. This perspective is crucial for understanding the long-term effects of environmental pollution, particularly when situated in the context of the intertwined relationships between historical legacies, power structures, and the ongoing activities of various actors in the network. These connections are equally pertinent to the exploration of the colonial and asymmetrical power structures manifested within the domain of fast fashion and the trading of low-quality garments at the Kantamanto Market. As posited by environmental humanities scholars Ilenia Iengo and Marco Armiero, the presence of toxic substances and narratives has the profound ability to accumulate within both human and more-than-human bodies and imaginaries, subtly shaping the nature of individual and collective existence.[64] Particularly concerning narratives that mirror societal disparities, injustices, and the interplay of asymmetrical power relations, the transformative journey of once-valued garments takes a poignant turn, evolving into mere commodities for buyers, only to culminate in what can be termed as toxic waste.[65] This transformation serves as a conduit, carrying with it the narratives of toxicity and socioenvironmental injustices, which span multiple temporal realms, as described by Iengo and Armiero: "the biological, the personal, and the collective, including both humans and more-than-humans."[66]

Herein, Nigerian-American writer, photographer, and art historian Teju Cole's definition of the "white savior complex" finds resonance, encapsulating the phenomenon where individuals from the Global North, predominantly of white origin, perceive a calling to engage in developmental, educational, or aid endeavors within countries of the Global South.[67] This context reintroduces the concept of "white saviorism," allowing donors to feel a sense of empowerment, seemingly contributing to environmental well-being and upholding ethical principles on a human scale. Amid this perspective, narratives of toxicity remain muted or altogether imperceptible, obscured by the portrayal of donations in Western countries as commendable acts sanctioned by commercial and industrial players and stakeholders. However, it is imperative to acknowledge the prevailing reality of an unjust world in which Western countries and nations incessantly produce, consume, and discard at the expense of others.

Toxic Fumes and Airborne Plastics

The preceding discussion on "toxic timescapes" offers insight into the gradual accumulation of hazardous substances within the environment.[68] These substances, once introduced, endure, affecting ecosystem, human, and more-than-human well-being over time and space. Pollution thus embodies a form of contamination defined by the interplay of origins, chance occurrences, and resulting effects.[69] Initiating a cascade of degradation and disruption within the balance of the environment, pollutants can be identified as substances introduced into the environment—encompassing materials, particles, biological contaminants, and microorganisms.[70] Significantly, while natural events can cause pollution, the prevailing undertone of the term implies an anthropogenic origin.[71] One manifestation of pollution is air pollution—a phenomenon that extends to the realm of the atmosphere and encompasses a critical facet of environmental degradation that is closely linked to human and more-than-human health.[72] Air pollution, which consists of gases, liquids, or solids released into the air, comes from a variety of sources, including fossil-fuel combustion, vehicle emissions, industrial activities, open burning of waste, and construction.[73] Certain pollutants, such as particulate matter (PM), carbon monoxide, ozone, nitrogen dioxide, and sulfur dioxide, pose health hazards and are of great concern. Alarmingly, World Health Organization (WHO) data shows that almost the entire global population (99 percent) is exposed to air that exceeds recommended quality standards, illustrating a disturbing picture of widespread exposure to pollutants. It is noteworthy that air pollution tends to be more prevalent in low- and middle-income countries. Thus, air pollution involves the alteration of the composition of indoor and/or outdoor air by various pollutants, which represents a deviation from the natural atmospheric equilibrium.[74]

Equally relevant is the phenomenon of open burning of waste, which is emerging as one of the most hazardous sources of toxic aerosols in the environment.[75] Beth Gardiner, an American environmental journalist, paints a vivid image of smoke ascending in the air: "The smoke rises from thousands of fires, hundreds of thousands. Millions. Soot. Black carbon. The same particles that sicken the body, now heading toward the sky. . . . The effects, in an atmosphere already tipping out of balance, powerful enough to cause clouds to form and to disperse them. To alter the weather."[76] Her words capture the gravity of the situation in which the substances that affect our health are rising into the sky, causing a cascade of consequences. As she describes, it becomes clear how unbalanced the natural world has become, illustrating the interconnectedness between anthropogenic extractivism and its impact on nature. This alarming reality is supported by World Bank projections indicating that global solid waste generation will escalate to 3.4 billion tons per year by 2050.[77] The conventional disposal paradigms of municipal waste management include practices such as collection, recycling, composting, landfilling, and incineration.[78] However, in less industrialized and economically stable countries, small-scale open burning or unregulated landfilling dominate as two commonly used solutions.[79] Approximately 41 percent of the world's annual waste is disposed of through open burning.[80]

Focusing on the Kantamanto Market in Ghana, a significant amount of discarded clothing meets its ultimate fate on open lands or in dumping sites.[81] The impact of this practice may not be immediately apparent, but the unregulated burning of these garments poses significant risks to people working in these environments, particularly in terms of air pollution.[82] The majority of these garments are made of synthetic materials such as acrylic, polyester, nylon, and viscose, which have a negative impact on the environment.[83] The production and processing of these materials also involves the use of chemicals, dyes, and plasticizers, making them a twofold threat: jeopardizing both environmental integrity and the well-being of those involved in their production, use, and disposal.[84] The burning of predominantly synthetic fiber–based clothing adds to the air pollution and negatively impacts human health by polluting the atmosphere with microplastic fibers and a mixture of toxic and noxious fumes.[85] This concern is further supported by the study conducted by environmental engineer Luís Fernando Amato-Lourenço and colleagues that underscores the presence of microplastics in the very air that we breathe.[86] Consequently, the act of inhaling these microplastics can potentially lead to their accumulation in human lung tissue.[87] It is important to note that the origins of these airborne microplastics are diverse, but microplastic fibers from a wide range of synthetic textiles are predominant. Workers exposed to these conditions are at increased risk of developing a variety of lung diseases.[88] Additionally, the exhaust gases resulting from burning and incinerating, including CO_2, intensify the multifaceted environmental and health risks that are inseparably linked with air pollution.[89] Weather phenomena, such as wind currents, can act as vectors to disperse these plastic fragments across the atmosphere, contributing to the dissemination of plastic particles in the airscapes.[90] Extended exposure to air pollution, including microplastics, has been correlated with respiratory, cardiovascular, and assorted health conditions.[91] In addition, the presence of plastic particles in the air exacerbates existing environmental concerns and increases ecological disruption within local ecosystems.

Amid the backdrop of the COVID-19 pandemic, air pollution has further implications for individual well-being and flourishing. Research demonstrates that air pollution presents a substantial risk factor contributing to adverse impacts on the cardiovascular and respiratory systems, including infections such as pneumonia and influenza.[92] Notably, particles suspended in the air can act as carriers for pathogens, thereby promoting the spread of diseases. This includes air pollutants such as $PM_{2.5}$ and PM_{10}, sulfur dioxide, nitrogen dioxide, carbon monoxide, and ozone, which can impair the respiratory system through inhalation and worsen susceptibility to respiratory virus infections.[93] Concerning the transmission of SARS-CoV-2, organic chemist Antonio Frontera and colleagues have proposed that a setting characterized by elevated levels of airborne pollutants, coupled with specific climatic circumstances, may enhance the period during which viral particles remain airborne.[94] Aligning with this, economist Mario Coccia underscores the role of specific air pollutants, notably PM_{10} or ozone, in shaping the transmission dynamics of COVID-19.[95] However, a deeper exploration of this topic is essential to understand the correlations between air pollutants and COVID-19 spread. Nonetheless, it is likely that fine particulate matter concentration is not the sole environmental factor involved in the spread of SARS-CoV-2. Various environmental factors such as size and composition of atmospheric particulate matter as well as meteorological conditions including wind velocity, humidity, and temperature play a role.[96] It can be said, however, that already marginalized communities residing in areas afflicted by elevated air pollution levels are even more vulnerable to respiratory illnesses.

The implications of these interactions extend far beyond the immediate context of the Kantamanto Market and the global secondhand textile industry as the complex links between pollution, plastic waste, and the secondhand clothing market are explored. The intertwining of environmental degradation, health concerns, and socioeconomic dynamics underscores the urgent need for comprehensive and sustainable solutions. In addressing the far-reaching effects of pollution from the uncontrolled incineration of discarded clothing, particularly those made of synthetic fibers, it is important to recognize the gravity of the challenges.

Conclusion

Driven by the unbridled need to produce and consume rapidly, the global fashion industry has dumped an estimated ninety-two million tons of clothing into landfills worldwide. A microcosm of the paradoxes of this industry is the Kantamanto Market, a bustling hub where millions of kilograms of discarded clothing from the Western world find new life. Here, used clothing is either resold or tossed into nearby dumpsites, where incineration produces toxic fumes and contributes to local air pollution and other environmental and health problems. Closer exploration of this nexus reveals a picture of interconnectedness, power dynamics, and systemic inequalities. Reflecting on the unjust and colonial narratives embedded in the secondhand clothing market transforms the meaning of clothing. It goes beyond being merely a means of expressing oneself or practicality, and becomes a symbol of historical injustice, bearing the imprint of domination and colonial influence. In this light, the notion of "fueling toxicity" takes on a particular sharpness. It describes the mechanism by which various actions, structures, and entities within the secondhand clothing trade amplify environmental and societal harms.

The theoretical framework on which this essay is based allows us to understand the intricacies and entanglements of this global industry. ANT reveals the stage on which networks of human and more-than-human entities interact in the patterns of production, consumption, and disposal. It is a multilayered account in which "waste regimes," in their hybridity, determine the fate of once-valued garments. Through the lens of ANT, waste can be seen as both a material reality and a cultural construct, a notion viewed through the prism of privilege and colonialism. This complex interplay of toxicity and colonialism finds definition in the concept of "toxic timescapes," where the "slow violence" of pollution intertwines with temporal patterns that encompass marginalized communities.[97] This is the perspective from which the Kantamanto Market presents itself, revealing the unequal burden of pollution suffered by those who are often forgotten in the narratives of the pasts, presents, and futures. Here, once-cherished garments are transformed into toxic substances, and narratives of white saviorism and alleged goodwill merge with histories of environmental injustice. Air, once a source of life and vitality, becomes a medium carrying the burden of waste, plastics, and chemical toxins. The fumes from incinerated garments contribute to the ongoing crisis of air pollution, affecting environmental systems and human and more-than-human well-being and flourishing.

In this discussion of intertwined issues related to the toxicity of fast fashion and the secondhand industry, one thing is particularly clear—it demands attention. This essay aims to point a finger at the ecological impacts of fast fashion, the trade dynamics perpetuating inequalities, and the colonial legacies, not to assign blame, but to foster a better understanding that many issues of the climate crisis are embedded in colonial power dynamics, hierarchical structures, historical narratives, and injustices that fuel toxicity. It is more about promoting mindfulness through awareness—a discussion that calls for reflection, consciousness, and collective engagement for change. The intertwined web of consequences resulting from unregulated and constant pollution and textile waste underscores the undeniable connection between our choices as consumers, the material life cycle, and the well-being of the environment. The lessons that can be drawn from the Kantamanto Market and the global secondhand clothing trade extend far beyond its immediate boundaries and reflect the shared responsibility that we all bear for the health of our planet and for justice in our societies.

NOTES

1. Xuandong Chen et al., "Circular Economy and Sustainability of the Clothing and Textile Industry," *Materials Circular Economy* 3, no. 12 (December 2021): 1–9, https://doi.org/10.1007/s42824-021-00026-2; Martina Igini, "10 Concerning Fast Fashion Waste Statistics," *Earth.org*, August 2, 2022, https://earth.org/statistics-about-fast-fashion-waste/.

2. Jo Lorenz, "Decolonising Fashion: How an Influx of 'Dead White Man's Clothes' Is Affecting Ghana," *Eco Age*, August 4, 2020, https://eco-age.com/resources/decolonising-fashion-dead-white-mans-clothes-ghana/.

3. "The Impact of Textile Production and Waste on the Environment (Infographics)," European Parliament, June 5, 2023, https://www.europarl.europa.eu/news/en/headlines/society/20201208STO93327/the-impact-of-textile-production-and-waste-on-the-environment-infographics.

4. Andrew Brooks, *Clothing Poverty: The Hidden World of Fast Fashion and Second-Hand Clothes* (London: Zed Books, 2015), 56–71; Zeynep Ozdamar Ertekin and Deniz Atik, "Sustainable Markets: Motivating Factors, Barriers, and Remedies for Mobilization of Slow Fashion," *Journal of Macromarketing* 35, no. 1 (March 2015): 53–69, https://doi.org/10.1177/0276146714535932.

5. Christopher Breward, *Fashion* (Oxford: Oxford University Press, 2003), 54; Brooks, *Clothing Poverty*, 56–71; "Was ist Polyester? Ein Einblick in den berüchtigten 'Love It or Hate It' Stoff," *Contrado* (blog), November 23, 2017, https://de.contrado.com/blog/was-ist-polyester/; Margret Nemak, "Nylon—eine Kunstfaser erobert die Welt," *Der Mitteldeutsche Rundfunk*, September 20, 2018, https://www.mdr.de/geschichte/ddr/wirtschaft/nylon-perlon-kunstfaser-dederon-100.html.

6. "Was ist Polyester?," *Contrado*.

7. Breward, *Fashion*, 23; Brooks, *Clothing Poverty*, 56–71; Alma Rominger, "The History & Rise of Fast Fashion: From the 18th Century to Today," GROWensemble, July 20, 2023, https://growensemble.com/history-of-fast-fashion/.

8. Breward, *Fashion*, 23; Brooks, *Clothing Poverty*, 56–71.

9. Ertekin and Atik, "Sustainable Markets"; Julian M. Allwood et al., "Well Dressed? The Present and Future Sustainability of Clothing and Textiles in the United Kingdom," University of Cambridge, Institute for Manufacturing, November 2006, https://www.ifm. eng.cam.ac.uk/uploads/Resources/Other_Reports/UK_textiles.pdf; Sonja Geiger, Samira Iran, and Martin Müller, "Nachhaltiger Kleiderkonsum in Dietenheim: Ergebnisse einer repräsentativen Umfrage zum Kleiderkonsum," University of Ulm, May 2017, https://doi. org/10.13140/RG.2.2.23686.98888.

10. Zsuzsa Gille, "Actor Networks, Modes of Production, and Waste Regimes: Reassembling the Macro-Social," *Environment and Planning A: Economy and Space* 42, no. 5 (May 2010): 1049–64, https://doi.org/10.1068/a42122; Zsuzsa Gille, *From the Cult of Waste to the Trash Heap of History: The Politics of Waste in Socialist and Postsocialist Hungary* (Bloomington: Indiana University Press, 2007); Zsuzsa Gille, "From Risk to Waste: Global Food Waste Regimes," *Sociological Review* 60, no. 2_suppl (December 2012): 27–46, https://doi.org/10.1111/1467-954X.12036.

11. Gille, "Actor Networks."

12. Ilenia Iengo and Marco Armiero, "Toxic Bios: Traversing Toxic Timescapes through Corporeal Storytelling," in *Toxic Timescapes: Examining Toxicity across Time and Space*, ed. Simone M. Müller and May-Brith Ohman Nielsen (Athens: Ohio University Press, 2023), 187–211.

13. Rob Nixon, *Slow Violence and the Environmentalism of the Poor* (Cambridge, MA: Harvard University Press, 2011).

14. Timothy W. Luke, "Reconstructing Social Theory and the Anthropocene," *European Journal of Social Theory* 20, no. 1 (February 2017): 80–94, https://doi. org/10.1177/1368431016647971; Matthew Lepori, "There Is No Anthropocene: Climate Change, Species-Talk, and Political Economy," *Telos*, no. 172 (September 2015): 103–24, https://doi.org/10.3817/0915172103; Scott Hamilton, "I Am Uncertain, but We Are Not: A New Subjectivity of the Anthropocene," *Review of International Studies* 45, no. 4 (October 2019): 607–26, https://doi.org/10.1017/S0260210519000135.

15. Luke, "Reconstructing Social Theory"; Lepori, "There Is No Anthropocene"; Hamilton, "I Am Uncertain, but We Are Not."

16. The OR Foundation (@theorispresent), "Kantamanto Delegation Reflections Part 3," Instagram, February 7, 2023, https://www.instagram.com/reel/CoWe44pNbIG/?utm_source=ig_web_copy_link&igshid=MzRlODBiNWFlZA==.

17. Gille, "Actor Networks."

18. Swatiprava Rath and Pranaya Kumar Swain, "The Interface between Political Ecology and Actor-Network Theory: Exploring the Reality of Waste," *Review of Development and*

Change 27, no. 2 (September 2022): 264–78, https://doi.org/10.1177/09722661221122553.

19. Eva Pongrácz, "Re-Defining the Concepts of Waste and Waste Management: Evolving the Theory of Waste Management" (PhD diss., University of Oulu, 2002), 69, http://urn.fi/urn:isbn:9514268210.

20. Martin O'Brien, *A Crisis of Waste? Understanding the Rubbish Society* (New York: Routledge, 2008), 26.

21. O'Brien, *A Crisis of Waste?*, 27.

22. Linda Valkeman, "1_Trailer_Sender_Receiver_Conversation," Vimeo, November 24, 2021, 2:17 to 3:10, https://vimeo.com/649455550?login=true#_=_. Video no longer available.

23. Marco Armiero and Massimo De Angelis, "Anthropocene: Victims, Narrators, and Revolutionaries," *South Atlantic Quarterly* 116, no. 2 (April 2017): 345–62, https://doi.org/10.1215/00382876-3829445; Marco Armiero, *Wasteocene: Stories from the Global Dump* (Cambridge: Cambridge University Press, 2021), https://doi.org/10.1017/9781108920322.

24. Gille, *From the Cult of Waste*; Gille, "Actor Networks"; Gille, "Risk to Waste."

25. Max Liboiron, *Pollution Is Colonialism* (Durham, NC: Duke University Press, 2021); Max Liboiron, "Modern Waste Is an Economic Strategy," *Discard Studies*, July 9, 2014, https://discardstudies.com/2014/07/09/modern-waste-is-an-economic-strategy/.

26. Gay Hawkins, "Disposability," *Discard Studies*, May 21, 2019, https://discardstudies.com/2019/05/21/disposability/; Gay Hawkins and Stephen Muecke, eds., *Culture and Waste: The Creation and Destruction of Value* (Lanham: Rowman & Littlefield, 2013).

27. Bruno Latour, *Reassembling the Social: An Introduction to Actor-Network-Theory* (Oxford: Oxford University Press, 2005); John Law, "Notes on the Theory of the Actor-Network: Ordering, Strategy, and Heterogeneity," *Systems Practice* 5, no. 4 (August 1992): 381–82; Rath and Swain, "Interface between Political Ecology," 269.

28. George Ritzer, *Encyclopedia of Social Theory* (Thousand Oaks: Sage Publications, 2005), 1–3.

29. Latour, *Reassembling the Social*; Martin Müller and Carolin Schurr, "Assemblage Thinking and Actor-Network Theory: Conjunctions, Disjunctions, Cross-Fertilisations," *Transactions of the Institute of British Geographers* 41, no. 3 (March 2016): 218, https://doi.org/10.1111/tran.12117.

30. Ignacio Farías and Sophie Mützel, "Culture and Actor Network Theory," in *International Encyclopedia for the Social and Behavioral Sciences*, ed. James D. Wright, 2nd ed. (Amsterdam: Elsevier, 2015), 525, http://dx.doi.org/10.1016/B978-0-08-097086-8.10448-9.

31. Müller and Schurr, "Assemblage Thinking and Actor-Network Theory," 218.

32. Bruno Latour, "On Actor-Network Theory: A Few Clarifications," *Soziale Welt* 47 (1996).

33. Latour, "On Actor-Network Theory," 371.

34. Latour, "On Actor-Network Theory," 373.

35. Latour, "On Actor-Network Theory," 373; Latour, *Reassembling the Social*, 46.

36. Gille, "Actor Networks," 1050.

37. Gille, "Actor Networks," 1050–51.

38. Gille, "Actor Networks," 1051.

39. Gille, "Actor Networks," 1053.

40. Gille, "Actor Networks," 1056.

41. Masters of Good/L. Delebois (@mastersofgood), "Is donating our clothes a good idea?" Instagram, April 20, 2021, https://www.instagram.com/p/CN5nu1_JsmE/.

42. Linda Lampel, "Value Capture and Distribution in Second-Hand Clothing Trade: The Role of Charity Discourses, Commercial Strategies and Economic and Political Contexts" (PhD diss., University of Vienna, 2020), https://www.oefse.at/fileadmin/content/Downloads/Publikationen/Foren/forum72_lampel_web.pdf.

43. Masters of Good/L. Delebois (@mastersofgood), "Is donating our clothes a good idea?"

44. Lucy Norris, "The Limits of Ethicality in International Markets: Imported Second-Hand Clothing in India," *Geoforum* 67 (December 2015): 183–93, https://doi.org/10.1016/j.geoforum.2015.06.003.

45. Mike Crang et al., "Rethinking Governance and Value in Commodity Chains through Global Recycling Networks," *Transactions of the Institute of British Geographers* 38, no. 1 (January 2013): 12–24, https://doi.org/10.1111/j.1475-5661.2012.00515.x.

46. J. Branson Skinner, "Fashioning Waste: Considering the Global and Local Impacts of the Secondhand Clothing Trade in Accra, Ghana and Charting an Inclusive Path Forward" (master's thesis, University of Cincinnati, 2019), 95–96, https://etd.ohiolink.edu/acprod/odb_etd/etd/r/1501/10?clear=10&p10_accession_num=ucin1553613566277155; Liz Ricketts and J. Branson Skinner, "Stop Waste Colonialism: Leveraging Extended Producer Responsibility to Catalyze a Justice-Led Circular Textiles Economy," OR Foundation, February 14, 2023, 8–9, https://stopwastecolonialism.org/stopwastecolonialism.pdf.

47. Joline d'Arnaud van Boeckholtz, "The Secondhand Clothing Supply Chain: Tracing Translations of Objects of Clothing from the Global North to Ghana" (master's thesis, Erasmus University Rotterdam, 2020), 4, https://thesis.eur.nl/pub/61382/M.-Thesis-Joline-d-Arnaud-van-Boeckholtz-571882.pdf; Skinner, "Fashioning Waste," 98.

48. Skinner, "Fashioning Waste," 5; Isaac Kaledzi, "Used Clothes Choke Ghana's Markets, Ecosystem," *Deutsche Welle*, May 1, 2022, https://www.dw.com/en/

used-clothes-choke-both-markets-and-environment-in-ghana/a-60340513.

49. Andrew Brooks and David Simon, "Unravelling the Relationships between Used-Clothing Imports and the Decline of African Clothing Industries: African Clothing Industries and Used-Clothing Imports," *Development and Change* 43, no. 6 (November 2012): 1265–90, https://doi.org/10.1111/j.1467-7660.2012.01797.x; Brooks, *Clothing Poverty*, 129–30, 145, 156–60.

50. The OR Foundation (@theorispresent), "Heavy Are the Clothes on Your Back—Part 1," Instagram, January 17, 2023, https://www.instagram.com/p/CKJZHETApOD/.

51. Linda Valkeman, "Dead White Men's Clothes," *Obroni Wawu*, https://obroniwawu.com; Bryan Benjamin, "The Secondhand Market at the Heart of Ghana's Fashion Revolution," *Dazed*, August 2, 2022, https://www.dazeddigital.com/fashion/article/56679/1/kantamanto-is-the-second-hand-market-fuelling-ghana-s-fashion-revolution; van Boeckholtz, "Secondhand Clothing Supply Chain," 24–26.

52. Benjamin, "The Secondhand Market"; van Boeckholtz, "Secondhand Clothing Supply Chain," 24–26.

53. Van Boeckholtz, "Secondhand Clothing Supply Chain," 24–26.

54. Brooks, *Clothing Poverty*, 31–50; van Boeckholtz, "Secondhand Clothing Supply Chain," 26; Kristina R. Anderson, Eric Knee, and Rasul Mowatt, "Leisure and the 'White-Savior Industrial Complex,'" *Journal of Leisure Research* 52, no. 5 (October 2021): 531–34, https://doi.org/10.1080/00222216.2020.1853490.

55. Brooks, *Clothing Poverty*, 31–38.

56. Simone M. Müller and May-Brith Ohman Nielsen, "Introduction," in Müller and Ohman Nielsen, *Toxic Timescapes*, 1–14.

57. Nixon, *Slow Violence*, 2.

58. Müller and Ohman Nielsen, "Introduction," 6–11.

59. Nixon, *Slow Violence*, 2.

60. Müller and Ohman Nielsen, "Introduction," 6–11; Nixon, *Slow Violence*, 2.

61. Thom Davies, "Slow Observation: Witnessing Long-Term Pollution and Environmental Racism in Cancer Alley," in Müller and Ohman Nielsen, *Toxic Timescapes*, 50–71.

62. Davies, "Slow Observation," 50.

63. Davies, "Slow Observation," 50–51.

64. Iengo and Armiero, "Toxic Bios," 191.

65. Iengo and Armiero, "Toxic Bios," 191; van Boeckholtz, "Secondhand Clothing Supply Chain," 24–26.

66. Iengo and Armiero, "Toxic Bios," 191; van Boeckholtz, "Secondhand Clothing Supply Chain," 5; Benjamin, "The Secondhand Market."

67. Teju Cole, "The White-Savior Industrial Complex," *The Atlantic*, March 21, 2012, https://www.theatlantic.com/international/archive/2012/03/the-white-savior-industrial-complex/254843/; Anderson, Knee, and Mowatt, "Leisure and the 'White-Savior Industrial Complex.'"

68. Müller and Ohman Nielsen, "Introduction," 6–11.

69. Tatiana Konrad, Chantelle Mitchell, and Savannah Schaufler, "Introduction: Toward a Cultural Axiology of Air," in *Imagining Air: Cultural Axiology and the Politics of Invisibility*, ed. Tatiana Konrad (Exeter: University of Exeter Press, 2023), 1–34.

70. Frank R. Spellman, *The Science of Environmental Pollution* (Boca Raton: CRC Press, 2021); Richard Fuller et al., "Pollution and Health: A Progress Update," *The Lancet Planetary Health* 6, no. 6 (June 2022): e535–47, https://doi.org/10.1016/S2542-5196(22)00090-0; Diane Boudreau et al., "Pollution," *National Geographic*, December 14, 2022, https://education.nationalgeographic.org/resource/pollution; Konrad, Mitchell, and Schaufler, "Toward a Cultural Axiology," 10.

71. Konrad, Mitchell, and Schaufler, "Toward a Cultural Axiology," 10.

72. Ioannis Manisalidis et al., "Environmental and Health Impacts of Air Pollution: A Review," *Frontiers in Public Health* 8 (February 2020): 9, https://doi.org/10.3389/fpubh.2020.00014; "Air Pollution," World Health Organization (WHO), 2022, https://www.who.int/health-topics/air-pollution#tab=tab_1; Awais Piracha and Muhammad Tariq Chaudhary, "Urban Air Pollution, Urban Heat Island and Human Health: A Review of the Literature," *Sustainability* 14, no. 15 (July 2022): 4–5, https://doi.org/10.3390/su14159234.

73. Manisalidis et al., "Environmental and Health Impacts," 5–7; "Air Pollution: How We're Changing the Air," UCAR Center for Science Education, https://scied.ucar.edu/learning-zone/air-quality/air-pollution; Rishu Agarwal et al., "Chemical Composition of Waste Burning Organic Aerosols at Landfill and Urban Sites in Delhi," *Atmospheric Pollution Research* 11, no. 3 (March 2020): 554, https://doi.org/10.1016/j.apr.2019.12.004.

74. "Air Pollution," WHO.

75. Agarwal et al., "Chemical Composition," 554.

76. Beth Gardiner, *Choked: Life and Breath in the Age of Air Pollution* (Chicago: University of Chicago Press, 2019), 123, https://doi.org/10.7208/chicago/9780226630793.001.0001.

77. Silpa Kaza et al., "What a Waste 2.0: A Global Snapshot of Solid Waste Management to 2050," Overview booklet, World Bank, Washington, DC, September 2018, 3, https://doi.org/10.1596/978-1-4648-1329-0; Kapil Dev Sharma and Siddharth Jain, "Municipal Solid Waste Generation, Composition, and Management: The Global Scenario," *Social Responsibility Journal* 16, no. 6 (June 2020): 918, https://doi.org/10.1108/

SRJ-06-2019-0210.

78. Christine Wiedinmyer, Robert J. Yokelson, and Brian K. Gullett, "Global Emissions of Trace Gases, Particulate Matter, and Hazardous Air Pollutants from Open Burning of Domestic Waste," *Environmental Science & Technology* 48, no. 16 (August 2014): 9523, https://doi.org/10.1021/es502250z.

79. Navarro Ferronato and Vincenzo Torretta, "Waste Mismanagement in Developing Countries: A Review of Global Issues," *International Journal of Environmental Research and Public Health* 16, no. 6 (March 2019): 1–28, https://doi.org/10.3390/ijerph16061060.

80. Wiedinmyer, Yokelson, and Gullett, "Global Emissions," 9525.

81. Ferronato and Torretta, "Waste Mismanagement," 1–2, 7–15.

82. "The Impact of Textile Production," European Parliament.

83. Cristina Palacios-Mateo, Yvonne Van Der Meer, and Gunnar Seide, "Analysis of the Polyester Clothing Value Chain to Identify Key Intervention Points for Sustainability," *Environmental Sciences Europe* 33, no. 1 (December 2021): 2, https://doi.org/10.1186/s12302-020-00447-x; Mathilde Charpail, "What's Wrong with the Fashion Industry?," *Sustain Your Style*, 2017, https://www.sustainyourstyle.org/en/whats-wrong-with-the-fashion-industry; Lucy Jones, "5 Fashion Materials You Didn't Realise Were Bad for Wildlife," *BBC Earth*, https://www.bbcearth.com/news/5-fashion-materials-you-didnt-realise-were-bad-for-wildlife.

84. Aravin Periyasamy, "Microfiber Emissions from Functionalized Textiles: Potential Threat for Human Health and Environmental Risks," *Toxics* 11, no. 5 (April 2023): 4–22, https://doi.org/10.3390/toxics11050406.

85. Dohee Kwon et al., "Valorization of Synthetic Textile Waste Using CO2 as a Raw Material in the Catalytic Pyrolysis Process," *Environmental Pollution* 268 (January 2021): 1–3, https://doi.org/10.1016/j.envpol.2020.115916.

86. Luís Fernando Amato-Lourenço et al., "An Emerging Class of Air Pollutants: Potential Effects of Microplastics to Respiratory Human Health?," *Science of the Total Environment* 749 (December 2020): 1–7, https://doi.org/10.1016/j.scitotenv.2020.141676.

87. Lauren C. Jenner et al., "Detection of Microplastics in Human Lung Tissue Using µFTIR Spectroscopy," *Science of the Total Environment* 831 (July 2022): 1–10, https://doi.org/10.1016/j.scitotenv.2022.154907; Amato-Lourenço et al., "Emerging Class of Air Pollutants."

88. Johnny Gasperi et al., "Microplastics in Air: Are We Breathing It In?," *Current Opinion in Environmental Science & Health* 1 (February 2018): 1–5, https://doi.org/10.1016/j.coesh.2017.10.002; Amato-Lourenço et al., "Emerging Class of Air Pollutants."

89. Manisalidis et al., "Environmental and Health Impacts."

90. Laura Revell, "Microplastics Are in the Air We Breathe and in Earth's Atmosphere, and They Affect the Climate," Greenpeace, October 21, 2021, https://www.greenpeace.org/aotearoa/story/microplastics-are-in-the-air-we-breathe-and-in-earths-atmosphere-and-they-affect-the-climate/.

91. Byeong-Jae Lee, Bumseok Kim, and Kyuhong Lee, "Air Pollution Exposure and Cardiovascular Disease," *Toxicological Research* 20, no. 2 (June 2014): 71–75, https://doi.org/10.5487/TR.2014.30.2.071; Manisalidis et al., "Environmental and Health Impacts."

92. José L. Domingo and Joaquim Rovira, "Effects of Air Pollutants on the Transmission and Severity of Respiratory Viral Infections," *Environmental Research* 187 (August 2020): 1–7, https://doi.org/10.1016/j.envres.2020.109650; Maryam Maleki et al., "An Updated Systematic Review on the Association between Atmospheric Particulate Matter Pollution and Prevalence of SARS-CoV-2," *Environmental Research* 195 (April 2021): 1–7, https://doi.org/10.1016/j.envres.2021.110898; Alexys Monoson et al., "Air Pollution and Respiratory Infections: The Past, Present, and Future," *Toxicological Sciences* 192, no. 1 (March 2023): 3–14, https://doi.org/10.1093/toxsci/kfad003.

93. Domingo and Rovira, "Effects of Air Pollutants," 4; Maleki et al., "Updated Systematic Review," 4.

94. Antonio Frontera et al., "Regional Air Pollution Persistence Links to COVID-19 Infection Zoning," *Journal of Infection* 81, no. 2 (August 2020): 318–56, https://doi.org/10.1016/j.jinf.2020.03.045.

95. Mario Coccia, "Factors Determining the Diffusion of COVID-19 and Suggested Strategy to Prevent Future Accelerated Viral Infectivity Similar to COVID," *Science of the Total Environment* 729 (August 2020): 1–20, https://doi.org/10.1016/j.scitotenv.2020.138474.

96. Maleki et al., "Updated Systematic Review," 6.

97. Nixon, *Slow Violence*, 2.

Part 2

Health, Sustainability, and Race

Environment and Health

The Impact of Historical Environment on Inuit Qanuinngitsiarutiksait in the Era of Anthropogenic Climate Change

Jeevan Stephanie Kaur Toor, Tagaaq Evaluardjuk-Palmer, and Josée G. Lavoie

Inuit culture is inseparable from the condition of their physical surroundings. . . .
Environmental upheaval resulting from climate change violates Inuit's right to practice
and enjoy their culture.
—Sheila Watt-Cloutier, *The Right to Be Cold*

There are roughly sixty-five thousand Inuit in Canada, with many in Inuit Nunangat, the northernmost part of Canada, including fifty-three communities in four regions: Inuvialuit, Nunavut, Nunavik, and Nunatsiavut; an increasing population lives in southern Canada.[1] Historically, Inuit have lived off the Land,[2] harboring strong relationships with the environment, living and nonliving entities that inhabit their surroundings. This sentiment is captured by the above quotation by Sheila Watt-Cloutier, an Inuk from Nunavik who is an advocate for Inuit social and environmental issues. However, recent generations of Inuit communities have undergone swift cultural changes due to interactions with explorers and more recently from the detrimental actions of the Canadian government (e.g., relocation, residential schools). Traditionally, Inuit have adapted to their environment; however, recent transitions have occurred too quickly.[3] For example, Inuit youth suicide rates are now ten times those of non-Indigenous counterparts.[4] While this is not solely attributed to climate change, it highlights the existence of detrimental impacts.

Exploration of Inuit Qaujimajatuqangit (Inuit knowledge) makes apparent reasons for strong connection to the environment. Briefly, it is related to a traditional way of life that is lived in symbiosis with the Land. However, impacts of colonialism on the historical environment directly affect qanuinngitsiarutiksait and now are being exacerbated by climate change. Thus, these issues need to be considered together. Historically, it is known that global temperatures fluctuate. "Anthropogenic climate change" refers to changes in weather that are more rapid and occur due to human action. Due to global warming patterns the Arctic is at the forefront of these changes, with temperatures rising at quadruple the global average rate.[5] Thawing of permafrost releases greenhouse gases such as methane and carbon dioxide, contributing to an enhanced change in climate.[6] For Inuit, climate change has led to unpredictable weather patterns, for example, more rain, thinner ice, and warmer temperatures.[7] Inuit are especially susceptible to impacts of changing climate due to their close relationship with the environment, which is an essential protective factor for qanuinngitsiarutiksait.

Globally, there are 350 million Indigenous individuals who have strong connections to their environments, undermined by their experience of colonialism.[8] This is not to homogenize Indigenous people around the world, or even in Inuit Nunangat; the literature points out that there are differences in culture, environment, and impacts of climate change. Rather, this demonstrates that this work has applications for other communities, and this perspective emphasizes the notion of the interconnected world in which we live.

Methodology

This exploration is based on a review of the literature, supported by a research team working on Inuit health in Winnipeg, connected to Elders and Inuit-based organizations. The three authors of this essay bring a wealth of experience and perspective. Jeevan Toor, a PhD student at University College London exploring issues of Inuit health and well-being, had the idea to explore the interaction between health and environment from an Inuit perspective due to conversations with the community around the importance of the environment for Inuit wellness and the need to address health from an Inuit perspective, rather than biomedical ones. Prior to starting her PhD, Jeevan worked as a research assistant with a team in Winnipeg (on "The Qanuinngitsiarutiksait Project"—joint research between the University of Manitoba and a community-based organization), which allowed her to situate this research with the appropriate individuals from the outset. Tagaaq Evaluardjuk-Palmer is an Inuit Elder who lived near Pond Inlet on a traditional camp, first moving to Manitoba in 1980. Tagaaq is part of the Inuit Elders Executive Council on the aforementioned project. The research topic was discussed with Tagaaq to ensure its relevance to the community, and then discussions around Inuit knowledge, concepts, and understandings provided the basis for this exploration. Jeevan and Tagaaq worked collaboratively, where Jeevan Toor presented each draft to Elder Tagaaq until they were both satisfied with the outcome. Jeevan then worked with Josée G. Lavoie, a professor at the Indigenous Institute of Health and Healing (University of Manitoba) and principal investigator of the above mentioned project, to gain her perspective on this exploration. This guidance was sought as Dr. Lavoie has worked with the Inuit community for many years and has gathered a wealth of experience in this setting.

Due to constraints in traveling, findings are not from fieldwork, so a semi-ethnographic approach was adopted. This was done through three aspects:

1. framing research through Inuit concepts gained from conversation with Elder Tagaaq,
2. a literature search on Google Scholar to form the data, and
3. framing of sections with Inuit poetry or quotations from conversation or the analyzed literature.

This approach was taken as it allows somewhat for the richness of the anthropological qualitative approach to emerge and aligns with the importance of oral formats in Inuit culture. Specifics of this semiethnographic data collection are explained below.

Literature from the last twenty years was found using Google Scholar, through the following search terms: "Inuit health and climate change," "Inuit environment and health," and "Inuit climate change adaptions." It was sorted thematically by manually highlighting relevant information on climatic impacts, climatic adaptations (relating to mental health, food and water insecurity, disease, ice-related accidents, persistent organic pollutants), and proposals to move forward (split into government, policy, and research).

It was anticipated that full poems from Inuit authors could be included to frame the various sections within the material; however, due to copyright issues, this was not possible. Instead, excerpts of these poems have been included, and they will be discussed in the sections they are included. Two excerpts are from the contemporary poet Taqralik Partridge who is from Nunavik. Partridge is a spoken-word poet, and the poetry had not been written down until recently. The collection of poems referenced here was published in 2020. Partridge is also an artist, writer, and curator taking inspiration from the surrounding world of Inuit issues and poses questions that should be considered by all. Original formatting of these poems was preserved to reflect how the poems would be if they were spoken. The other poem is from a collection of Inuit poetry collected by Knud Rasmussen (Greenlandic-Danish explorer, 1879–1933) and translated by Tom Lowenstein in 1973. While the included poetry is from the past and present, similarities can be found in the poets' connection to the environment.

Inuit Qaujimajatuqangit, Being Healthy, and the Environment

Inuit Qaujimajatuqangit (IQ) translates as "that which Inuit have known all along," encompassing Inuit worldview through stories, teachings, and myths linking the past and future together, a "living' knowledge."[9] Often, IQ has been sidelined by Western research, as it was not recognized as "scientific" enough, that instead focused on biomedical approaches.[10] IQ is metaphysically different from Western knowledge; therefore comprehension in the context of Inuit culture, history, and beliefs is needed.[11] At the basis of IQ is Inuit maligait, four "laws" by which Inuit live: "Working for the common good and not being motivated by personal interest or gain, living in respectful relationships with every person and thing that one encounters, maintaining harmony and balance, planning and preparing for the future."[12] Cultural laws and communal laws also exist as an extension of these principles.[13] A foundational belief of IQ rests in a profound relationship with nature and the connection it provides to their environment.

For example, animals were not driven to extinction when Inuit beliefs prevailed; when Western influence predominated muskox and whales were hunted to near extinction.[14] This idea of connection to the surrounding environment is captured in the excerpt from Taqralik Partridge, "i can tell you how we sat low in canoes / or slept curved against the hull of a peterhead / on long trips up or down the river/ or out into the bay."[15] The excerpt is from the poem titled "after an argument" where memories of a childhood spent sleeping on trips across the water are explored. Due to interconnected beliefs, qanuinngitsiarutiksait cannot be understood without IQ; to fully conceptualize qanuinngitsiarutiksait, deeper insight into the Inuk is needed.[16]

Inuit Notion of the Person

Pool and Geissler argue that personhood is understood culturally in various ways; personhood is not an artifact of the brain but is one of "life and experience."[17] Personhood is a dialogue between the projection of society onto the individual and the individual forming their location in society. More fitting for this context is the concept "complex personhood,"[18] allowing emphasis on more than the individual; for example, family, community, the ancestors, and the environment become more prominent (ecocentrism). This allows a shift from Western emphasis on individuality to understanding a more complete notion of the person, which is closer to Inuit philosophy.[19]

Thus, Inuit conceptions of personhood, belonging, and identity lie in their strong relationship with the environment. If this relationship is lost, Inuit are left with a weakened sense of health and well-being, consequently weakened qanuinngitsiarutiksait. While connection to the Land is key in maintaining qanuinngitsiarutiksait, and therefore Inuk personhood, other factors are also integral: for example, cultural practices such as eating specific foods, speaking one's language, and having a strong sense of community.[20] Gombay suggests that colonialism created "colonised identities" as local norms defining aspects of personhood conflicted with colonizers' views of being and colonizers viewed Inuit as "uncivilised," lacking the qualities and benefits of Western personhood and consequently lacking citizenship.[21] Colonialism, past and present, continues to disrupt Inuit personhood and consequently identity and belonging. Attributes maintaining Inuit personhood, identity, and belonging have been impacted by colonialism, disrupting an Inuk's core personhood, and continue to be: for example, the modern biopolitical approach to Inuit, as explored by Stevenson.[22] This colonial disruption to identity and by extension personhood is further impacted by anthropogenic climate change. These disruptions impact Inuit qanuinngitsiarutiksait. Negative climatic impacts on the environment could possibly be triggering an environmental trauma response rooted in colonial treatment of the environment, now being brought to the surface because of climate change. The literature has begun to explore the disconnect between Inuit and the environment, due to climatic impacts, but does not yet link historical and wider environmental aspects, which we consider imperative for understanding the basis of these issues. Further work is needed to explore the suggestion of an environmental trauma response, as this was out of the scope of this research.

Through ethnography, d'Anglure unravels Inuit oral history, mythology, and cosmology to understand Inuit notions of being and belonging in the world and how these are at the heart of Inuit identity.[23] These findings reinforce the idea that Inuit qanuinngitsiarutiksait is affected by relationships with the world around oneself. Although the relationships occur from and to an Inuk, this does not mean a person should be the center of the world, but rather that one should live in balance with the cosmos; reinforcing that in this context of health and well-being is intimately connected to the larger environment.

Qanuinngitsiarutiksait

The use of the concept qanuinngitsiarutiksait pays homage to the cultural context of this exploration and acknowledges that Inuit concepts of good health differ from biomedical concepts and that past explorations have not embraced Inuit concepts. Insights reported here are based on conversations with Elder Evaluardjuk-Palmer. Qanuinngitsiarutiksait roughly translates as "good health and well-being," referring to a holistic all-encompassing idea of health, for example, well-being of the household, health of the person or Land. The main facets around qanuinngitsiarutiksait can be broken down into physical health, mental health, as well as cultural, spiritual, and environmental connection. It should be noted that these categories are somewhat arbitrary, but a necessary breakdown for those unfamiliar with Inuit knowledge. This concept also refers to tools used for health, for example, eating traditional foods, knowledge of traditional language, community connection, being on the Land, and having a strong sense of identity. Elder Evaluardjuk-Palmer conveyed the sentiment that research needs to involve Inuit in the process, to ensure a broad understanding, rather than just investigating an issue or aspect. Another Elder, Taamusi Qumaq, suggested that good health and well-being means to "live without worry, being able to move on the land with ease . . . hunt animals and eat the food they provide, and visiting and taking pleasure in the company of family and loved ones,"[24] reinforcing that qanuinngitsiarutiksait encompasses factors of one's larger social setting, such as family, nature, and community.

In contrast, biomedical ideas of good health and well-being are based upon Descartes's concept of mind-body dualism; now there is the suggestion that health and well-being are more of a multifaceted experience.[25] This nuance is demonstrated as the World Health Organization defines health as "a state of optimal physical, mental, and social well-being, and not merely the absence of disease and infirmity";[26] however, this does not take into account that the "social" aspect of health varies in different contexts. Mehta takes this distinction further, suggesting that physicians are unaware of the historical groundings of biomedical practice, therefore causing further harm in treatments.[27] The biomedical view does not allow for connection to cultural identity and spiritual health to be included in such definitions, and therefore in treatment. Acknowledging that disconnects exist between Inuit knowledge and Western approaches allows recognition of more than one way of knowing or understanding the world, integral for work aiming to situate itself in a decolonized realm, allowing production of "legitimate knowledge."[28] In a sense, biomedical beliefs are actually cultural constructions themselves, and the legitimization of projecting these beliefs onto others perpetuates colonist knowledge.

Environment, Avatittingnik Kamatsiarniq

Although the environment is part of the "Inuit determinants of health," it is mostly disregarded in non-Indigenous contexts.[29] Importance of the environment is emphasized in the quotation by Elder Evaluardjuk-Palmer, where she suggests the idea of environmental stewardship to underpin Inuit attitudes toward the environment: "We lived on the land, off the land, for the land . . . What we took we put back . . . My parents only took what they needed and lived for today."[30] Some anthropologists and geographers have engaged with this concept as an avenue to guide health-related research, with anthropologists specifically grappling with ways in which power affects interactions between humans and their environment.[31] Geographers suggest that the environment "refers to the sum total of conditions which surround man at a given point in space and time."[32] The subdiscipline of health geography suggests that the concept of "landscape" should be used to explore health-environment interactions as it encompasses "diverse and converging layers of history, social structure, and built environment at particular sites."[33] These perspectives are helpful to begin to comprehend interactions between health and environment.

As highlighted by anthropologists and geographers, Land is an important concept to grapple with in health-related research. Also, as explored above, Land is an integral part of maintaining qanuinngitsiarutiksait. Indigenous authors have described the Land as "everything: the identity, the connection to our ancestors, the home of our nonhuman kinfolk, our pharmacy, our library, the source of all that sustains us," where the soul and the Land and people are likened to be part of an interconnected web with no hierarchy between people and the environment.[34] These understandings are held by Inuit, as health and well-being are positively impacted by being on the Land and carrying out associated activities such as hunting, fishing, going to family cabins.[35] Connection to the Land is described by Inuit as "defining who we are" and that "people are not only from a particular place, but they are also of the place, that is their identities, well-being, livelihoods, histories and emotion-spiritual connections are emergent from the lands on which they live."[36]

Thus, environment (and by extension Land) is a facet of the Inuk experience. Niewöhner and Lock theoretically encapsulate the connection between environment and the person from an anthropological position through the concept "situated biologies," suggesting that the environment can influence "environment/human entanglements."[37] This concept emerged from the term "local biologies" that suggests health is impacted by the material body. Using "situated" instead of "local" implies that it is not just personal biological makeup that impacts the individual but also the environment. Further, the term "situated" acknowledges that not only local factors impact the body, which is increasingly apparent in our global world. This builds upon Haraway's "situating knowledges," that it is social, political, historical, and economic aspects that structure and restructure the body; they are in constant flux, allowing bodies to be in a state of "becoming."[38] This ontological caveat is wholly applicable to help to unpack relationships between qanuinngitsiarutiksait and the environment.

Perhaps more fitting is the Inuktitut concept of "avatittingnik kamatsiarniq," which can be seen to some degree to combine perspectives explored above. This concept was shared by Elder Evaluardjuk-Palmer who stressed the importance of understanding environment from an Inuit perspective. Avatittingnik kamatsiarniq roughly means an all-encompassing situation of someone in their surroundings, where one is aware of the habits of the Land and acts toward it with care and respect. The person is not the most prominent feature in the Land, but rather part of an interconnected web of its surroundings. Elder Evaluardjuk-Palmer noted that other aspects such as policy, government, and research are now part of avatittingnik kamatsiarniq, hence their inclusion later. For Inuit, one is connected to and part of the cosmos; thus comprehension of historical context is necessary to understand issues of adverse qanuinngitsiarutiksait.

Historical Environment

Colonialism refers to imperialist actions based on capitalist expansionist ideals, where the dominant society utilizes their power to attain benefits.[39] The following excerpt is from Taqralik Partridge's poem "Colonisation Is a Pyramid Scheme": "Next, you'll say they're one of you. The ones that made it this far. / Not the top of the pyramid but close. Close enough to get the gold / watch. Close enough to lead the pep rally. Close enough that you / can see yourself in their movements. And now your job is done."[40] In this poem, the process of colonialization toward Inuit is explored with witty remarks and is likened to a pyramid scheme. It is suggested that even if an Inuk had followed the Western way of life, they would have never quite got to the "top" of the pyramid, but that it would have been alright as it would have been deemed a Western success. This comparison could be likened to capitalism and the hierarchy of society; indoctrination into a wage-based economy, reliance on welfare and Western foods, and assimilation into a different way of life are also covered. Hinting toward the creation of Inuit stereotypes in relation to colonial and subsequent racist ideology is mentioned, too, as can be understood through, for example, the "lazy Inuk."

Inuit first experienced interactions with Europeans from explorers and whalers in the sixteenth century; in the latter half of the twentieth century the Canadian government introduced policies affecting Inuit way of life through sedenterization, forced relocation (some distances similar to Toronto to Miami), implementation of a monetary system, and residential schools.[41] Residential schools impacted children through loss of connection to culture, language, and community, with the last school for Inuit closing in 1996.[42] Forced relocation ensured Inuit no longer lived their nomadic lifestyle, with families moved to overcrowded matchbox housing with few facilities; mortgages had to be paid, but monetary jobs were uncommon. Thus, indoctrination began into a Western way of life and a dependency on social welfare.[43] The Royal Canadian Mounted Police killed 1,200 sled dogs (used for hunting), officially to stop spreading disease, but implicitly to "encourage" Inuit to stay in settlements, deepening their reliance on government handouts as hunting for food then became difficult. Actions such as this have been attributed to a policy of assimilation for gain of Canadian territorial rights in the high Arctic.[44]

Colonialism set the foundations for capitalism, an ideology based on monetary growth leading to industrialization and consequently climate change.[45] Davis and Todd link colonial actions to broken connections between mind, body, and Land.[46] The concept of an "ecological paradox" applies, as at first mistreatment of the environment leads to economic growth, but then this growth comes at the cost of health and the environment.[47] Note that positive impacts from this growth are only experienced in certain places, and negative impacts from this growth are not experienced equally, that is, temperatures rising to a higher degree in the Arctic and subsequential impacts. Thus, anthropogenic climate change links to impacts to the avatittingnik kamatsiarniq, rooted in colonial actions, negatively impacting qanuinngitsiarutiksait. Recent literature found that aspects of wellness, for example, kinship and culture, were negatively affected by welfare/settler colonialism and that there were intersections between the effects of colonialism and aspects of coping with climate issues.[48] Thus colonialism, and subsequent welfare/settler colonialism, is responsible for the impact of and Inuit response to climate change and for anthropogenic climate change, ultimately impacting Inuit qanuinngitsiarutiksait. Using anthropological and geographical theory as a foundation, built upon with Inuktitut concepts, a holistic understanding of qanuinngitsiarutiksait comes to fruition that gains distance from biomedical models of health. Without interdisciplinarity that wholly considers Inuit knowledge, we cannot fully comprehend these problems at hand.

Climatic Impacts on Qanuinngitsiarutiksait

Literature suggests that the main impacts of climate change on qanuinngit-siarutiksait are related to mental health, food insecurity, rise of diseases, water insecurity, ice-related accidents, and persistent organic pollutants. For clarity, these issues will be investigated separately, but their interconnection should be noted. Interconnection of climate-related impacts are highlighted by a domino analogy:

> It's all the pieces, like dominoes, all touches each other. I mean everything you do, [our] Inuit way of life and our way of thinking is all intertwined and interconnected [to the environment]. So, something as significant as changes in the temperature, and in snow and rain and that kind of thing, it's all going to have a ripple effect.[49]

It is important to highlight that literature does not mention any benefits from these climatic changes, exemplified by an Inuit hunter stating, "the weather has changed, and has changed for the worse."[50] The idea of experiencing joy from the elements of the environment is explored in one of Rasmussen's poems, titled "Moved," collected from an Iglulik Inuk woman named Uvavnuk: "The sky's height stirs me. / The strong wind blows through my mind . . . carries me with it . . . I shake with joy."[51] Given this historical account of a positive emotion associated with being on the Land and aligning with more recent perspectives, the importance of an Inuk being in the physical environment is seen to be a visceral connection.

Adaptations to climate change will be discussed, demonstrating resilience, centering Inuit voices and initiatives, moving from deficit-orientated research; however, as many issues are relatively new, adaptations do not always exist.

Mental Health

Spending less time on the Land has led to feelings of anxiety, depression, anger, and frustration. Changes to ice reliability, which is becoming thinner or where open water remains for longer periods of time, are due to warming temperatures, resulting in "feeling trapped" or "like a caged animal."[52] Durkalec and colleagues found interviewees saying if they could not go onto the Land they would "have no health," "can't breathe," "be lost," and "their appetite and mind would go."[53] In Nunatsiavut, visits to the health clinic for mental health issues increased after a period of warmer weather, and when temperatures cooled visits were less frequent;[54] one health practitioner noted, "when people can't get out . . . we see a difference in the counselling part: people are more agitated . . . because they're just not getting off on the land."[55] Literature noted spending less time on the Land negatively impacted Inuit sense of identity through being unable to do culturally based activities and engage with traditional knowledge,[56] impacting feelings of self-worth and productivity. This is exemplified by a hunter who said:

> If . . . people can't be going to the cabin . . . hunting and . . . going on the land, then . . . [we] start to see a community shifting, not knowing what they're supposed to be doing . . . not knowing what your self-worth is, not knowing what you should be doing with your time.[57]

Studies found that spending less time on the Land has led to increased ideations of suicide and use of drugs and alcohol. In Rigolet, an interviewee stated, "people get bored . . . turn to drinking and drugging."[58] Community cohesion has been impacted by spending less time on the Land, impacting mental wellness, with an interviewee stating, "cohesiveness is usually a common denominator of outdoor-based activities . . . it's sort of somewhat fragmented [because of climate change]. The cohesiveness that now bonds the community could be jeopardized because what else are you bonding on?"[59]

Not spending time outside means more time at home, which some found stressful due to many people in a small space: "people felt like they were getting in each other's way more, in a way that they previously hadn't experienced."[60] Lastly, studies suggested less time on the Land increased attention to historic trauma:

If you are able to find some sense of worth in yourself . . . able to start unconsciously healing from those wounds . . . if for some reason you are not able to do something that makes you feel good . . . then those tragedies . . . [are] still there, and then they're magnified because they come more to the surface because you're not feeling personal strength.[61]

However, when more time is spent on the Land, interviewees expressed feeling "more relaxed, calm and peaceful, as well as healthier and happier."[62] One interviewee encapsulated the benefits of spending time outside as "much a part of our life as breathing . . . so if we don't get out then . . . it's like taking part of your arm away . . . you're not fulfilled. . . . It's just like taking medicine."[63]

Access to mental health services in Inuit Nunangat is inconsistent, as Western-trained practitioners generally come to communities for a limited time. Or people rely on telehealth, and practitioners may know little of the culture.[64] While telehealth services can be useful, for example, during the COVID-19 pandemic, it poses issues for those without reliable Internet (inaccessible to many),[65] or for those who feel more comfortable with in-person conversations. Solutions could include more Inuit training in health-care programs, located in the north, that are Inuit-centric, or training practitioners extensively in the cultures in which they are working. Practitioners need to be aware of the impacts of changing Land on mental health, which could be included in their training.[66] Drug and alcohol abuse require a holistic approach, addressing all facets of health.[67] Land-based programming, successful in nurturing youth wellness, should continue.[68]

Food Insecurity

Traditional "country" foods benefit Inuit economically, culturally, and nutritionally, nourishing mental and spiritual health and community cohesion through hunting, harvesting together, or sharing food according to prescribed rules of mutual care.[69] This sentiment is demonstrated by one interviewee: "I love that [country] food, it's really healthy . . . it's good for your body and your spirit. . . . You feel good about going out on the Land and being able to do that."[70] However, climatic changes have altered vegetation growth, migration patterns and ocean acidification levels, melting ice coinciding with seal births, which leads to high mortality, caribou preferring areas with more snow in winter and cooler areas in summer, thinner caribou due to less vegetation, caribou drowning when crossing thinner ice, and fewer berries found with less flavor.[71] Changes exacerbate existing issues of food security; a 2018 publication shows food insecurity rates in the Arctic were eight to ten times higher than the rest of Canada, and data from 2008 show 70 percent of houses in Nunavut have dealt with food security issues—the highest rate globally for an Indigenous populace in an advanced country.[72]

Watt-Cloutier relates Inuit's preference for country foods: "country food connect[s] us to water, land, to the 'source' of our life," and not eating them is a "spiritual loss." Hunting teaches valuable skills such as "patience, boldness, tenacity, focus, courage, sound judgement and wisdom, very transferable to the modern world that has come so quickly."[73] Ostapchuk and colleagues explained the link between food and personhood, through identity, as follows: "cultural identity is partly what you eat . . . part of identifying with being Inuit is eating [country foods]."[74] Positive health benefits from traditional food are noted through intake of essential vitamins, minerals, and antioxidants, with less refined carbohydrates, saturated fat, and sodium.[75] General well-being through consumption of traditional foods has meant people "feel warmer, [it] relaxes the body, causes less stress, makes them feel more full, causes less bloating."[76] Kirmayer and colleagues found general feelings of discontent when country foods were not eaten, such as "weakness, lassitude and tiredness . . . irritability, uncooperativeness . . . depression." Another said, "Inuit eat mainly meat because it has blood in it and that helps . . . the person [to] be in better health . . . It's visible even on the cheeks, the cheeks were redder . . . in the past."[77]

Consumption of store-bought foods has increased due to the demand of the market economy, where the nine-to-five work week, Monday to Friday, interferes with harvesting activities that depend on the weather; the loss of skills associated with residential schools and weather changes also impact harvesting and hunting. Purchased foods are twice as expensive or more as in the rest of Canada.[78] Healthy items are available but are even more expensive, leading to higher consumption of processed foods.[79] This dietary shift, along with spending less time on the Land, has been linked to higher noncommunicable conditions such as diabetes, high blood pressure, obesity, and cardiovascular issues.[80] An interviewee stated:

People are eating processed meat and salt and additives . . . the rate of diabetes has jumped really high. . . . Thirty years ago . . . I do not think there were any diabetics in town. Now there're lots," with another stating, "there's a higher increase of obesity . . . heart disease . . . high blood pressure and heart attacks and strokes and I think that it is related to food.[81]

Adaptation has been demonstrated in Arviat through use of community greenhouses to grow fresh produce, using seaweed as fertilizer, and community freezers to store and share successful hunts.[82] In Kugluktuk hunters have revised their courses to reflect changes in migratory patterns, and Elder-youth mentorship initiatives have been implemented to help convey traditional knowledge of hunting practices that can then be adapted to specific climate needs.[83] However, further investment needs to be put toward resolutions and strategies to lower costs of food.

Diseases

More rain, less snow, and warmer temperatures have contributed to spread of disease in the Arctic.[84] A parasite from the Amazon, *Toxoplasma gondii* (causing flu, vision issues, neurological problems, and stillbirth, and passed to fetuses in vitro), is spread either by felids traveling to the Arctic from oocysts discharged in water via north-traveling currents, infecting fish and entering the Arctic aquatic food chain, or from migrating animals infected in the Amazon region. Both hypotheses are caused by climate change, as oocysts reach the Arctic from changing currents due to warming temperatures, or migratory animals have changed routes due to climate changes.[85] Further, increased Arctic water levels have allowed gastrointestinal pathogens, particularly *Helicobacter pylori* and *Campylobacter*, to be transmitted more in recent years causing gastritis, ulcers, and gastric cancer.[86] In Cambridge Bay and Kugluktuk, these pathogens were found in muskox, making them inedible.[87] Lack of infrastructure in water-testing systems has led to underreporting of these diseases, as many pathogens are transported in water.[88] Increase of these diseases is associated with climate change, therefore testing capacities need to increase, and practitioners need an up-to-date knowledge of rates, treatment, and mitigation measures that are culturally appropriate.

Research into modes of transmission and the impact these pathogens have on these communities needs further investigation.

Water Insecurity

Most Canadians have access to drinking water, but not in Inuit Nunangat, as not all communities receive consistent water distribution.[89] Communities receive water via water tanks or aged pipe structures from the 1950s, which leak, causing lower pressure and wasted water. Unreliability of clean water means boil water advisories are frequent; trucked water is dependent on deliveries that are often unreliable, directly contravening the United Nation's human rights resolution on access to clean water.[90] The government prefers trucked water, due to low construction and maintenance costs, but running costs are high.[91] As with food insecurity, climatic changes exacerbate existing issues of water insecurity. In a previously unknown event in Iqaluit in 2019 water pipes froze, due to changes in temperature, and the city spent $33,000 to thaw pipes, which took two weeks, leaving those who relied on this distribution without water.[92] Those using piped water (often coming from nearby lakes or reservoirs) find provision unreliable due to warming temperatures, which melts permafrost under water sources leading to evaporation; in Hopedale in 2015 the school and health clinic closed due to low water reserves.[93] Traditionally water was collected from the Land, but sources of water are drying up. As noted: "There are a lot more brooks that dried up. And there are a lot more ponds that are drying. I notice when I go out on the land . . . to where we used to get water maybe twenty-five years ago . . . the brooks there are really dried up now."[94] Preference of the traditional practice of collecting water from the Land has been expressed, with one person saying they would collect water from brooks during Land-based activities, then take it back to the community, and another saying:

> I don't drink [tap water] unless it's an emergency—I would drink a glass then, if I had no water here and the store was closed and I couldn't get up to the brook. Well, then I'd sip on a little bit. Mostly if I have to use that water, I'll boil it first.[95]

The literature notes that water availability is expected to vary via season, but in recent summers it has been harder to collect water.[96] An interviewee stated:

I started noticing about five years ago. Out around our cabin where we go in the summertime what used to be ponds are now mud holes [and] when there is less water it's closer to the ground so it might be boggier and dirty. . . . If you have ample water supply you'll get it from a running brook which will be healthier.[97]

Due to an unreliable and inaccessible water supply, attributed to infrastructure and climatic issues, many rely on buying bottled water, which is extremely expensive; some have filtered water systems, but this too is becoming unaffordable. Further, rising gas prices have been identified as an issue for those who prefer to collect water from the Land.[98] Preference for Land water must be recognized as valid, and the drying up of natural sources of water needs to be addressed. Until then, filtered and bottled water must be economically and readily accessible, with investment for better infrastructure.

Ice-Related Accidents

Ice is essential for traveling; climate change has led to thinner ice and a new problem of less predictability of which ice is safe, as traditional knowledge was normally sufficient.[99] Transmission of traditional knowledge has been negatively impacted by residential schools and by Inuit's need to engage with a diversity of activities that compete with their time, such as school and wage employment. Rates of injury and death while traveling on the Land have increased, and search and rescue incidents have increased.[100] This issue is exemplified by the following account:

My neighbour . . . had fallen through the ice on a hunting trip; . . . he pulled himself out of the water and, in soaking wet clothes, dug himself into a snowdrift for insulation; . . . he spent nearly two days that way before he was found. Both his legs were already frozen, so they had to be amputated.[101]

Loss of ability to travel on the Land impacts mental wellness, food security, and connection to culture as well as safety.

Adaptation to this new reality has been shown by technological initiatives. For example, synthetic aperture radar remote sensing, combined with traditional knowledge, has been developed to identify safeness of ice. Cambridge Bay hunters have found this useful and suggest it could be developed into maps and used in schools to teach children how to go out on the Land.[102] Another initiative, SmartICE, uses a similar idea and works in collaboration with Inuit to provide this service.[103] Due to Internet and reception issues, experienced frequently in the region, maps should be downloaded before travel and available on lower bandwidths.[104]

Persistent Organic Pollutants

The impact of persistent organic pollutants (POPs) on qanuinngitsiarutiksait is another issue affecting Inuit. POPs originate from harmful manufactured chemicals, such as plastic and polychlorinated biphenyls, that impact the nervous, reproductive, and immune systems.[105] These pollutants have also been found to cause cancer and birth defects and often act as disruptors to the endocrine system.[106] Even though the Arctic does not produce these chemicals, it is more susceptible to them due to the grasshopper effect, as the chemicals evaporate in warmer airs in the southern hemisphere and come to colder regions in the form of rain or snow, acting as a chemical sink for anthropogenic substances.[107] POPs are then transmitted to Inuit through consumption of aquatic animals, through the process of bioaccumulation, as pollutants are soluble in fat (e.g., seal and whale blubber).[108] These chemicals can transfer from mother to infant through breastfeeding and from the spread of the pollutants across the placenta.[109] Watt-Cloutier recalls that when members of her community began questioning if they should give up breastfeeding or eating country foods, "the threat to our country food struck [her] at a deeply visceral, emotional level."[110] She highlights this problem was not "about politics, but rather about families, parents, children and grandchildren—and our right to lead our lives and continue the strong traditions of our hunting culture."[111]

During the 1980s this topic was much discussed, as for the first time the connection from these pollutants was made; after great advocacy, the Stockholm Convention was implemented in 2001 to combat use of these contaminants. Following this legislation there was a decline in these chemicals in the Arctic; now, increased temperatures are melting ice and thus releasing trapped pollutants into the atmosphere.[112] Once again Inuit will have to question whether the foods of their homeland are safe to consume, as stated by Watt-Cloutier in the 1980s:

The last thing we need at this time is to worry about the very country food that nourishes us, spiritually and emotionally, poisoning us. . . . This is not just about contaminants on our plate. This is about a whole way of being, a whole cultural heritage that is at stake. . . . [The] process of hunting and fishing, followed by the sharing of food, . . . is a time-honoured ritual that binds us together and links us with our ancestors.[113]

Concentrations of POPs in Inuit mothers' breastmilk are now higher than those in southern Canada. These pollutants contribute to diabetes, respiratory infections, and decreased weight and height in infants.[114] It has also been suggested that exposure to these pollutants can affect rates of cardiovascular disease due to higher rates of cholesterol.[115] However, this could also be due to higher reliance on store-bought food items and lower rates of exercise, as explored earlier. While the impacts of this issue are seen more significantly in the Arctic, lack of attention to mitigating POPs still affects the planet as a whole as effects further afield are not presenting yet. Attention to the interconnecting aspect of this issue was highlighted by Watt-Cloutier: "an issue that concerned the survival of Arctic peoples, and indeed the health of all people around the world. A poisoned Inuk child, a poisoned Artic, and a poisoned planet are all one and the same."[116]

Solutions need to be found to stop the release of these trapped chemicals into the atmosphere or to break down the chemicals, coping strategies implemented at the receiving end of their impacts, and a ban of future use of these chemicals. Due to pollutants' lengthy half-lives, it will take numerous years for the environment to recover, and often the harm of various chemicals is not known until after they are released back into the atmosphere when the negative impacts are seen—intervention at this point is too late.

The Arctic Council has paid less attention to this in recent years, and the climax of this campaign was the signing of the Stockholm Convention. This lack of attention is compounded by the United States still declining to ratify the convention.[117] However, the Inuit Circumpolar Council is still working on this issue in partnership with other international governing bodies such as the United Nations.[118] In 2023 the United Nations announced a treaty to be the first "global plastic pollution treaty."[119] Actions such as these are the starting point for addressing the issue of POPs again.

However, policy is only one side of this issue; Boyd and Furgal have highlighted that the ways in which the impacts of POPs are communicated to Inuit communities needs to be well thought out.[120] For example, there is no Inuktitut phrase that equates to "chemical contaminant." In the past those conveying information about this issue have often not placed emphasis on the benefits (e.g., cultural, spiritual, general wellness) of eating country foods, while just relaying the complex scientific message, leading to confusion. Sometimes there has been lack of trust in those communicating these messages, due to the suspicion of undermining of Inuit culture stemming from colonial legacies. Therefore, communication of POPs needs to be created in collaboration with Inuit. Further, studies relating to POPs did not tend to consider other determinants of health that could contribute to poorer health outcomes in infants, such as nutritional deficits and respiratory illnesses; while it is acknowledged that this is a complex issue, all facets should still be addressed to gain a complete picture of the challenges presented.

Where Do We Go from Here?

Inuit have been adapting to the avatittingnik kamatsiarniq since time immemorial; now with human-induced climatic changes, compounded by lasting colonial impacts, the historical environment needs to be addressed. Wider issues of government approach, policy legislation, and research methodology in this sphere need to be reconsidered as systematic problems feed into the wider discussion of climate change and qanuinngitsiarutiksait.

Government

The first Indigenous minster of justice and attorney general, Jody Wilson-Raybould, noted that overall government practice was to not pay adequate attention to Indigenous issues. Wilson-Raybould states that "government practice for generations was to deny, delay, distract when it came to Indigenous issues."[121] She comments that the Trudeau administration had strong rhetoric with little substance, a focus on keeping the party in power, a lack of understanding of what decolonization, reconciliation, and a nation-to-nation relationship means in practice, a lack of willpower to focus on Indigenous issues, an approach of "we know better" in relation to Indigenous issues, and allowing "politics of exclusion to exist." She sums up this administration as being "focused on power and partisanship and so little interested in principle . . . that was more image that substance," and points out that this government can make significant changes that are enacted quickly in situations of crisis—as shown through the COVID-19 response. However, the "government does not see the dire reality that many Indigenous peoples live in as a crisis."[122] Often, government attention toward the Arctic is on economic development through resource extraction, even though this does not allow for sustainable maintenance of the environment, and historically, mining projects have failed to implement maintainable economic improvement in Inuit Nunangat.[123]

In the context of climate change, the "Pan-Canadian Framework on Clean Growth and Climate Change" and "A Healthy Environment and a Healthy Economy" are governmental reports addressing issues of health and climate change.[124] While Indigenous people, including Inuit, were referred to throughout, they were not directly involved and did not provide insight into these recommendations.[125] Trudeau's visit to Nunavut was two years after he became prime minister, reflecting the lack of support this region receives.[126] Now, focus is on the Arctic in relation to climate change only because melting ice makes the nation more accessible to foreign threats, mirroring an attitude that saw relocation of Inuit in the 1950s—suggesting possible military presence in the Arctic again.[127] Watt-Cloutier highlights the perpetual misunderstanding in politics of the importance of climate change in the Arctic, exemplified by the Canadian delegation's refusal to include the Arctic in discussion at the Conference of the Parties in 2003, even though she was trying to highlight that the Arctic is the first place to show changes in climate, and that without action the rest of the world would follow.[128] This attitude, although from a government over twenty years ago, represents a government genuinely uninterested in a nation-to-nation relationship, and missing the urgency of understanding the climate crisis.

To fully understand problems Inuit face in this context, federal and provincial government needs this on their agenda, but before policy is designed and support given they must ensure that they have an in-depth understanding of IQ, Inuit personhood, and implications of the colonized identity. Impacts of climate change on qanuinngitsiarutiksait are interconnected and multifaceted, and it could be argued that Inuit are more vulnerable to these impacts due to underlying socioeconomic inequalities, such as high food prices, poor housing, water provision, and facilities to keep up with the rise of diseases. Lack of attention to these basic provisions contributes to systemic racism, or "environmental racism,"[129] and helps to perpetuate settler colonialism. Wilson-Raybould suggests the administration has taken the stance that the right to self-determination, and therefore self-government, is negotiable; arguably self-determination is the first step toward decolonization, reconciliation, and a nation-to-nation relationship.[130]

Policy

To remedy issues of qanuinngitsiarutiksait in the context of climate change, policy needs to be situated in a decolonized framework where structures in southern Canada are not copied, but rather adapted to particularities of those living in the Arctic; where revival of Inuit governance occurs through insights from the community and collaboration takes place between the community and those enacting policy. There is also a sense of urgency to acting on this matter as policy and large-scale change take time to implement. Watt-Cloutier only hopes that by taking action now toward climatic impacts that we have a chance of reversing the harms done to our planet: "at this late date our job is to build movements, ones powerful enough to force the policy changes that give us our only hope of catching up with physics."[131]

Inuit Tapiriit Kanatami (ITK), the representative body for Inuit in Canada, suggest collaboration between Inuit and those with access to contemporary climate data, allowing for a coordinated response influenced by IQ.[132] This approach has begun to be used, for example, SmartICE or the greenhouse initiative, but more needs to be done to fully encapsulate this approach. Alfred and Corntassel propose that for change to come about, "shifts in thinking and action" need to occur.[133] Shifts include, but are not limited to, revisiting the 1984 Health Act, greater focus on IQ, and qualitative data.

For context, the 1876 Indian Act does not directly apply to Inuit, but to other Indigenous groups in Canada, but the attitude of the Canadian government toward Inuit is rooted in these thoughts. Lack of direct mention to Inuit in this act further exemplifies how this community has been disregarded. In regard to treatment of Indigenous people in Canada Wilson-Raybould describes the existence of this act as "segregationist, colonial and racist," suggesting that it is because of "ignorance, fear, greed, and lack of will" that it still exists, and if it continues to exist "there will always be institutionalised, systematic racism in Canada with respect to Indigenous people."[134] The Canada Health Act is the bedrock of health policy, with its objective to "protect, promote and restore the physical and mental well-being of residents of Canada and to facilitate reasonable access to health services without financial or other barriers."[135] However, federal commitments to enacting policy assisting in health and well-being issues of Inuit is "limited" as there are no specific requirements that define "federal obligations towards Indigenous peoples."[136] Those in the Arctic often receive poorer standards of health care informed by southern models—as seen from treatment of respiratory syncytial virus in infants. Ultimately to achieve equity and maintenance of qanuinngitsiarutiksait these key pieces of legislation need to be revisited where IQ and Inuit voices inform practice.

Policy appears to be driven by economic quantitative data; however, it can be difficult to quantify issues such as the emotional response to climate change and the importance of Inuit knowledge and perspective.[137] Thus, for issues such as this attention needs to focus on qualitative data, such as ethnographic works that allow profound findings to surface, or adaptation of quantitative measures to be wholly applicable to this context. For example, ethnographic works from d'Anglure and Stevenson give insights on Inuit life, traditions, and issues that would not be evident from quantitative data. Comprehension to this level of detail is integral for the enactment of policy and legislation.

Researchers

Inuit have experienced a complicated history with researchers due to lack of consultation and consent, attributed to an approach based on colonial hierarchies.[138] Over the past few years research in Inuit Nunangat has become more of a partnership, for example, links being made between academics and Inuit with projects tackling priorities in the community.[139] Although a participatory approach has been used for research, often this is limited to consultation, and findings are not communicated back to the community.[140] The short and simple quotation of "nothing about us, without us" encapsulates the collaborative way in which research should be approached.[141]

ITK suggest five areas for productive and just research to take place: "advance Inuit governance in research; enhance ethical conduct of research; align funding with Inuit research priorities; ensure Inuit access, ownership, and control over data and information; build capacity in Inuit Nunangat research."[142] Further, they suggest there are issues with funding processes for Inuit-related research, as most funds are given to those in southern Canada or outside Canada, and that Inuit are limited in the funds they can access due to measures put in place by the federal government.

Methodologically and theoretically, a shift is developing in this field, and researchers should take it into consideration. For example, the way Liboiron footnotes and cites Indigenous scholars and the way in which "Land" is capitalized when it indicates a primary relationship (rather than being used in a general sense) should be considered, as used here, given the importance of acknowledging sources and connection to the Land.[143] Further, a shift is needed in how topics related to Inuit are labeled and discussed, as previous literature tends to focus on what has gone wrong rather than what is going well and the ways in which self-determination can be attained. Tuck suggests that rather than thinking of Indigenous people as broken or damaged, a fuller representation of the community should be taken into account, based on resilience, taking a strengths-based approach, allowing a shift from deficit-oriented research,[144] an approach attempted here.

Conclusion

Improving Inuit qanuinngitsiarutiksait requires holistic self-determination; this comes from acknowledging that colonial legacies still affect Inuit today, that approaches to research, government, and policy need to be reconsidered from a strengths-based approach, with acknowledgment of existing problems. This can be done by working in collaboration with Inuit, acknowledging and understanding Inuit Qaujimajatuqangit, recognizing that issues are interconnected and should not be viewed in isolation, taking into consideration the importance of the Land and its associated benefits, taking a cross-disciplinary perspective, not separating the mind and body in approach, acknowledging the value of qualitative approaches, and ultimately, commitment from the government for improvement. Future research needs to be located in qanuinngitsiarutiksait, Inuit knowledge, and Inuit methodologies and be Inuit-led; lack of attention to these facets can be seen to perpetuate ignorance and continuation of the colonial mindset. Additionally, researchers should aim to think in a cross-disciplinary manner, allowing for nuances and greater insights to come to light. While these considerations should be taken into account, a paradigm shift in approach and thinking is needed for substantive positive change to happen. Jody Wilson-Raybould says that "change takes courage, including the courage to break away from the old ways of doing things that are not achieving the needed results."[145]

Finding ways to adapt to current climatic realities through innovative solutions is a noble endeavor and may provide solutions to other issues too (e.g., the greenhouse initiative helping with rising costs of foods in the Arctic and SmartICE assisting with transgenerational teachings of the Land); the aim should not be to accept a worsening climate, but rather to commit to reversing anthropogenic climatic impacts so that Inuit and other Arctic Indigenous groups have "the right to be cold."[146] Exploration has also demonstrated that the impacts of climate change are compounded by existing socioeconomic disparities and the long-lasting impacts of colonialism. To be clear, this is not to say that Inuit in urban settings cannot achieve qanuinngitsiarutiksait, but the scope of this study is to investigate climatic impacts in the Arctic; further research should engage with the impact of climate change and the ways in which this narrative impacts urban Inuit.

Overall, the issue is not just about Inuit in the Arctic; it is about recognizing that the impacts of climate change occur in the Arctic first, that Inuit have been impacted from climate change earlier and to a greater extent than the rest of the globe. They have served as an early warning beacon, and our humanity has been lacking. Now, as citizens of an interconnected globe, acting for Inuit clearly equates to acting in our own interest, as well as assisting those who have been affected most acutely.

NOTES

1. Inuit Tapiriit Kanatami, "National Inuit Strategy on Research" (report, Ottawa, 2018), 1–44.
2. Capitalized to follow Liboiron's example emphasizing importance of the Land in this context, not used in quotations to keep original meaning.
3. Laurence Kirmayer, Caroline Tait, and Cori Simpson, "The Mental Health of Aboriginal Peoples in Canada: Transformations of Identity and Community," in *Healing Traditions: The Mental Health of Aboriginal Peoples in Canada*, ed. Laurence Kirmayer and Gail Valaskakis (Toronto: UBC Press, 2009), 3–35.
4. Polina Anang et al., "Building on Strengths in Naujaat: The Process of Engaging Inuit

Youth in Suicide Prevention," *International Journal of Circumpolar Health* 78, no. 2 (2019). https://doi.org/10.1080/22423982.2018.1508321.

5. Kaylia Little, "Iqaluit's Water Crisis Highlights Deeper Issues with Arctic Infrastructure," The Arctic Institute, May 2, 2022, https://www.thearcticinstitute.org/iqaluits-water-crisis-highlights-highlights-deeper-issues-arctic-infrastructure/; C. S. Mena, M. Artz, and C. Llanten, "Climate Change and Global Health: A Medical Anthropology Perspective," *Perspectives in Public Health* 140, no. 4 (2020): 196–97; Jacqueline Middleton et al., "Temperature and Place Associations with Inuit Mental Health in the Context of Climate Change," *Environmental Research* 198 (July 2021): 1–11, https://doi.org/10.1016/j.envres.2021.111166.

6. Carrie Arnold, "Toxic Trouble as the Arctic Heats Up," Chemical & Engineering News, August 27, 2023, https://cen.acs.org/environment/climate-change/Toxic-trouble-Arctic-heats/101/i28.

7. Laurence Lebel et al., "Climate Change and Indigenous Mental Health in the Circumpolar North: A Systematic Review to Inform Clinical Practice," *Transcultural Psychiatry* 59, no. 3 (2022): 312–36; Joshua Ostapchuk et al., "Exploring Elders' and Seniors' Perceptions of How Climate Change Is Impacting Health and Well-being in Rigolet, Nunatsiavut," *International Journal of Indigenous Health* 9 (2015): 6–24.

8. Taiaiake Alfred and Jeff Corntassel, "Being Indigenous: Resurgences against Contemporary Colonialism," *Government and Opposition* 40, no. 4 (2005): 597–614; Agata Durkalec et al., "Climate Change Influences on Environment as a Determinant of Indigenous Health: Relationships to Place, Sea Ice, and Health in an Inuit Community," *Social Science & Medicine* 136–37 (2015): 17–26.

9. Mark Kalluak, "About Inuit Qaujimajatuqangit," in *Inuit Qaujimajatuqangit: What Inuit Have Always Known to Be True*, ed. Joe Karetak, Frank Tester, and Shirley Tagalik (Winnipeg: Fernwood Publishing, 2017), 41; Shirley Tagalik, "Inuit Qaujimajatuqangit: The Role of Indigenous Knowledge in Supporting Wellness in Inuit Communities in Nunavut" (report, British Colombia, 2009), 1–8.

10. Priscilla Ferrazzi et al., "*Aajiiqatigiingniq*: An Inuit Consensus Methodology in Qualitative Health Research," *International Journal of Qualitative Methods* 18 (2019), https://doi.org/10.1177/1609406919894796; Katherine Wilson et al., "Changing the Role of Non-Indigenous Research Partners in Practice to Support Inuit Self-Determination in Research," *Arctic Science* 6, no. 3 (2020): 127–53.

11. Joe Karetak and Frank Tester, "Inuit Qaujimajatuqangit, Truth and Reconciliation," in Karetak, Tester, and Tagalik, *Inuit Qaujimajatuqangit*, 1–19.

12. Karetak and Tester, "Inuit Qaujimajatuqangit, Truth and Reconciliation," 3.

13. Leah McDonnell et al., "Unforeseen Benefits: Outcomes of the Qanuinngitsiarutiksait Study," *International Journal of Circumpolar Health* 81 (2022), https://doi.org/10.1080/22 423982.2021.2008614.

14. Karetak and Tester, "Inuit Qaujimajatuqangit, Truth and Reconciliation."

15. Partridge, *curved against the hull of a peterhead* (Guelph: PS Guelph, 2020), 236. 17; Extracts of this poem have been included to due space constraints, see full poem with reference.

16. Inuit means "the people"; Inuk is the singular of Inuit.

17. Robert Pool and Wenzel Geissler, *Medical Anthropology* (Berkshire: Open University Press, 2005), 117.

18. Lauren Dixon, "Complex Personhood," *Odd Thoughts* (blog), March 2, 2017, https:// deviantdixon.wordpress.com/2018/05/08/complex-personhood/.

19. Kirmayer, Tait, and Simpson, "The Mental Health of Aboriginal Peoples in Canada."

20. Ashlee Cunsolo Willox et al., "The Land Enriches the Soul: On Climatic and Environmental Change, Affect, and Emotional Health and Well-being in Rigolet, Nunatsiavut, Canada," *Emotion, Space and Society* 6 (2013): 14–24.

21. Nicole Gombay, "There Are Mentalities that Need Changing: Constructing Personhood, Formulating Citizenship, and Performing Subjectivities on a Settler Colonial Frontier," *Political Geography* 48 (2015): 11–23.

22. Lisa Stevenson, *Life beside Itself: Imagining Care in the Arctic* (Oakland: University of California Press, 2014).

23. Bernard Saladin d'Anglure, *Inuit Stories of Being and Rebirth: Gender, Shamanism, and the Third Sex* (Manitoba: University of Manitoba Press, 2018).

24. Christopher Fletcher et al., "Definition of an Inuit Cultural Model and Social Determinants of Health for Nunavik" (survey, Montreal, 2017).

25. Neeta Mehta, "Mind-Body Dualism: A Critique from a Health Perspective," *Mens Sana Monographs* 9, no. 1 (2011): 202–9.

26. "Health and Well-Being," World Health Organization, last modified September 12, 2023, https://www.who.int/data/gho/data/major-themes/health-and-well-being#:~:text=The%20WHO%20constitution%20states%3A%20%22Health,of%20 mental%20disorders%20or%20disabilities.

27. Mehta, "Mind-Body Dualism."

28. Max Liboiron, *Pollution Is Colonialism* (Durham, NC: Duke University Press, 2021), 21.

29. Inuit Tapiriit Kanatami, "Social Determinants of Inuit Health in Canada" (report, Ottawa, 2014), 1–44; Sherilee L. Harper et al., "Climate-Sensitive Health Priorities in Nunatsiavut, Canada," *BMC Public Health* 15, no. 1 (2015): 1–18, https://doi.org/10.1186/s12889-015-1874-3.

30. Tagaaq Evaluardjuk-Palmer (Inuit Elder) in discussion with J. Toor, February 6, 2023.

31. Hilda Kurtz and Karen Smoyer-Tomic, "Environment and Health," and Kevin Martin and Marianna Pavlovskaya, "Ethnography," both in *A Companion to Environmental Geography*, ed. Noel Castree, David Demeritt, Diana Liverman, and Bruce Rhoads (Chichester: John Wiley & Sons, 2009), 567–79 and 370–84, respectively; Luisa Cortesi, "Environmental Anthropology," in *The International Encyclopedia of Anthropology*, ed. Hilary Callan (Chichester: John Wiley & Sons, 2018), 3–19.

32. Chris Park, *Ecology and Environmental Management: A Geographical Perspective* (New York: Routledge, 1980), 29.

33. Robin Kearns and Damian Collins, "Health Geography," in *A Companion to Health and Medical Geography*, ed. Tim Brown, Sara McLafferty, and Graham Moon (Chichester: John Wiley & Sons, 2010), 17.

34. Robin Kimmerer, *Braiding Sweetgrass: Indigenous Wisdom, Scientific Knowledge and the Teachings of Plants* (London: Penguin, 2020), 50; Gregory Cajete, *Native Science* (Santa Fe: Clear Light Publishers, 1999); Enrique Salmón, *Eating the Landscape: American Indian Stories of Food, Identity, and Resilience* (Tucson: University of Arizona Press, 2012).

35. Willox et al., "The Land Enriches the Soul."

36. Ashlee Cunsolo Willox et al., "From This Place and of This Place: Climate Change, Sense of Place, and Health in Nunatsiavut, Canada," *Social Science & Medicine* 75, no. 3 (2012): 542.

37. Jörg Niewöhner and Margaret Lock, "Situating Local Biologies: Anthropological Perspectives on Environment/Human Entanglements," *BioSocieties* 13 (2018): 1–17.

38. Donna Haraway, *Simians, Cyborgs, and Women: The Reinvention of Nature* (New York: Routledge, 1991).

39. Karetak and Tester, "Inuit Qaujimajatuqangit, Truth and Reconciliation."

40. Partridge, *curved against the hull of a peterhead* (Guelph: PS Guelph, 2020), 45.

41. Kirmayer, Tait, and Simpson, "The Mental Health of Aboriginal Peoples in Canada."

42. Karetak and Tester, "Inuit Qaujimajatuqangit, Truth and Reconciliation"; Karen Restoule, "An Overview of the Indian Residential School System" (report, Ontario, 2013), 1–8.

43. Stevenson, *Life beside Itself.*

44. Sheila Watt-Cloutier, *The Right to Be Cold* (Minnesota: University of Minnesota Press, 2018); Stevenson, *Life beside Itself.*

45. Linda Connor, "Anthropogenic Climate Change and Cultural Crisis: An Anthropological Perspective," *Australian Journal of Political Economy* 66 (2010): 247–67; Christina Torrealba, "From Inuit Nunangat to the Marsh: How Climate Change and Environmental Racism Affect Population Health," *Healthy Populations Journal* 1, no. 2 (2021): 10–20; Kyle Whyte, "Indigenous Climate Change Studies: Indigenizing Futures, Decolonizing

the Anthropocene," *English Language Notes* 55 (2017): 153–62.

46. Heather Davis and Zoe Todd, "On the Importance of a Date, or, Decolonizing the Anthropocene," *International Journal for Critical Geographies* 16 (2017): 761–80.

47. Mena, Artz, and Llanten, "Climate Change and Global Health"; Middleton et al., "Temperature and Place Associations with Inuit Mental Health in the Context of Climate Change."

48. Cameron Mackay, "The Impact of Welfare Colonialism on Inuit Responses to Climate Change in Qikiqtani, Canada" (master's thesis, Cambridge University, 2018), 1–60; Whyte, "Indigenous Climate Change Studies."

49. Harper et al., "Climate-Sensitive Health Priorities in Nunatsiavut," 6.

50. Ostapchuk et al., "Exploring Elders' and Seniors' Perceptions," 14.

51. Tom Lowenstein, trans., *Eskimo Poems from Greenland and Canada* (London: Allison and Busby, 1973), 27.

52. Ostapchuk et al., "Exploring Elders' and Seniors' Perceptions"; Lebel et al., "Climate Change and Indigenous Mental Health in the Circumpolar North," 324; Willox et al., "The Land Enriches the Soul," 15.

53. Durkalec et al., "Climate Change Influences on Environment as a Determinant of Indigenous Health," 21.

54. Middleton et al., "Temperature and Place Associations with Inuit Mental Health in the Context of Climate Change."

55. Jacqueline Middleton et al., "'We're People of the Snow': Weather, Climate Change, and Inuit Mental Wellness," *Social Science & Medicine* 262 (2020): 6.

56. Susan Clayton, "Climate Anxiety: Psychological Responses to Climate Change," *Journal of Anxiety Disorders* 74 (2020), https://doi.org/10.1016/j.janxdis.2020.102263; Gabrielle Richards et al., "Commentary—The Climate Change and Health Adaptation Program: Indigenous Climate Leaders' Championing Adaptation Effort," *Health Promotion Chronic Disease Prevention in Canada* 39, no. 4 (2019): 127–30.

57. Willox et al., "From This Place and of This Place," 544.

58. Ostapchuk et al., "Exploring Elders' and Seniors' Perceptions," 17.

59. Harper et al., "Climate-Sensitive Health Priorities in Nunatsiavut," 12.

60. Willox et al., "The Land Enriches the Soul," 15.

61. Willox et al., "The Land Enriches the Soul," 16.

62. Lebel et al., "Climate Change and Indigenous Mental Health in the Circumpolar North," 321.

63. Willox et al., "The Land Enriches the Soul," 261.

64. Middleton et al., "Temperature and Place Associations with Inuit Mental Health in the

Context of Climate Change."

65. Inuit Tapiriit Kanatami, "The Digital Divide: Broadband Connectivity in Inuit Nunangat" (report, Ottawa, 2021), 1–16.

66. Clayton, "Climate Anxiety."

67. Watt-Cloutier, *The Right to Be Cold.*

68. C. Hackett et al., "Going Off, Growing Strong: Building Resilience of Indigenous Youth," *Canadian Journal of Community Mental Health* 35, no. 2 (2016): 79–82.

69. Watt-Cloutier, *The Right to Be Cold.*

70. Durkalec et al., "Climate Change Influences on Environment as a Determinant of Indigenous Health," 23.

71. Lebel et al., "Climate Change and Indigenous Mental Health in the Circumpolar North"; Harper et al., "Climate-Sensitive Health Priorities in Nunatsiavut"; Bindu Panikkar and Benjamin Lemmond, "Being on Land and Sea in Troubled Times: Climate Change and Food Sovereignty in Nunavut," *Land* 9, no. 12 (2020): 508.

72. Rebecca Schiff and Victoria Schembri, "Food and Health: Food Security, Food Systems and Health in Northern Canada," in *Health and Health Care in Northern Canada*, ed. Rebecca Schiff and Helle Møller (Toronto: University of Toronto Press, 2021), 40–56; Sarah Newell, "Social, Cultural, and Ecological Systems' Influence on Community Health and Wellbeing" (PhD thesis, McMaster University, 2018), 1–173.

73. Watt-Cloutier, *The Right to Be Cold*, 137, 202, 254.

74. Ostapchuk et al., "Exploring Elders' and Seniors' Perceptions," 17.

75. Noémie Boulanger-Lapointe et al., "Berry Plants and Berry Picking in Inuit Nunangat: Traditions in a Changing Socio-Ecological Landscape," *Human Ecology* 47 (2019): 81–93; Amy Caughey et al., "Niqivut Silalu Asijjipalliajuq: Building a Community-Led Food Sovereignty and Climate Change Research Program in Nunavut, Canada," *Nutrients* 14, no. 8 (2022): 1–12, https://doi.org/10.3390/nu14081572; Renata Rosol, Stephanie Powell-Hellyer, and Laurie Chan, "Impacts of Decline Harvest of Country Food on Nutrient Intake among Inuit in Arctic Canada: Impact of Climate Change and Possible Adaptation Plan," *International Journal of Circumpolar Health* 75, no. 1 (2016), https://doi.org/10.3402/ijch.v75.31127.

76. Newell, "Social, Cultural, and Ecological Systems' Influence on Community Health and Wellbeing," 25.

77. Laurence Kirmayer et al., "Inuit Concepts of Mental Illness: An Ethnographic Study" (report, Montreal, 1994), 54, 60.

78. Ashley Hayward et al., "A Review of Health and Wellness Studies Involving Inuit of Manitoba and Nunavut," *International Journal of Circumpolar Health* 79 (2020), https://

doi.org/10.1080/22423982.2020.1779524.

79. Panikkar and Lemmond, "Being on Land and Sea in Troubled Times."

80. Fatima Ahmed, Aleksandra Zuk, and Leonard Tsuji, "The Impact of Land-Based Physical Activity Interventions on Self-Reported Health and Well-Being of Indigenous Adults: A Systematic Review," *International Journal of Environmental Research and Public Health* 18, no. 13 (2021): 1–23; Schiff and Schembri, "Food and Health."

81. Willox et al., "From This Place and of This Place," 543.

82. Richards et al., "Commentary—The Climate Change and Health Adaptation Program"; Schiff and Schembri, "Food and Health."

83. Panikkar and Lemmond, "Being on Land and Sea in Troubled Times"; Rosol, Powell-Hellyer, and Chan, "Impacts of Decline Harvest of Country Food on Nutrient Intake among Inuit in Arctic Canada."

84. Richards et al., "Commentary—The Climate Change and Health Adaptation Program."

85. S. J. Reiling and B. R. Dixon, "Toxoplasma Gondii: How an Amazonian Parasite Became an Inuit Health Issue," *Canada Community Disease Report* 45, no. 7 (2019): 183–90.

86. Reiling and Dixon, "Toxoplasma Gondii"; Ahmed, Zuk, and Tsuji, "The Impact of Land-Based Physical Activity Interventions on Self-Reported Health and Well-Being of Indigenous Adults."

87. Panikkar and Lemmond, "Being on Land and Sea in Troubled Times."

88. Emma Finlayson-Trick et al., "Climate Change and Enteric Infections in the Canadian Arctic: Do We Know What's on the Horizon?" *Gastrointestinal Disorders* 3, no. 3 (2021): 113–26.

89. Inuit Tapiriit Kanatami, "Access to Drinking Water in Inuit Nunangat" (report, Ottawa, 2020), 1–17.

90. Inuit Tapiriit Kanatami, "Access to Drinking Water in Inuit Nunangat"; "About Water and Sanitation," United Nations, last modified April 28, 2023, https://www.ohchr.org/en/water-and-sanitation/about-water-and-sanitation#:~:text=OHCHR%20and%20the%20rights%20to%20water%20and%20sanitation,-Overview&text=On%2028%20July%20 2010%2C%20the,RES%2F64%2F292.

91. Little, "Iqaluit's Water Crisis Highlights Deeper Issues with Arctic Infrastructure."

92. Little, "Iqaluit's Water Crisis Highlights Deeper Issues with Arctic Infrastructure."

93. Inuit Tapiriit Kanatami, "Access to Drinking Water in Inuit Nunangat."

94. Christina Goldhar, Trevor Bell, and Johanna Wolf, "Vulnerability to Freshwater Changes in the Inuit Settlement Region of Nunatsiavut, Labrador: A Case Study from Rigolet," *Arctic* 67, no. 1 (2014): 76.

95. Goldhar, Bell, and Wolf, "Vulnerability to Freshwater Changes in the Inuit Settlement

Region of Nunatsiavut, Labrador," 77.

96. Harper et al., "Climate-Sensitive Health Priorities in Nunatsiavut"; Amy Kipp et al.,
"At-a-glance—Climate Change Impacts on Health and Wellbeing in Rural and Remote
Regions across Canada: A Synthesis of the Literature," *Health Promotion Chronic Disease
Prevention Canada* 39, no. 4 (2019): 122–26; Goldhar, Bell, and Wolf, "Vulnerability to
Freshwater Changes in the Inuit Settlement Region of Nunatsiavut, Labrador."

97. Goldhar, Bell, and Wolf, "Vulnerability to Freshwater Changes in the Inuit Settlement
Region of Nunatsiavut, Labrador," 76.

98. Goldhar, Bell, and Wolf, "Vulnerability to Freshwater Changes in the Inuit Settlement
Region of Nunatsiavut, Labrador."

99. Wilson et al., "Changing the Role of Non-Indigenous Research Partners in Practice to
Support Inuit Self-Determination in Research."

100. Rebecca A. Segal et al., "The Best of Both Worlds: Connecting Remote Sensing and Arctic
Communities for Safe Sea Ice Travel," *Arctic* 73, no. 4 (2020): 461–84; Torrealba, "From
Inuit Nunangat to the Marsh."

101. Watt-Cloutier, *The Right to Be Cold*, 186.

102. Segal et al., "The Best of Both Worlds."

103. SmartICE, "Enabling Resiliency in the Face of Climate Change," last modified July 10,
2022, https://smartice.org.

104. Segal et al., "The Best of Both Worlds."

105. Jan Dusik, "Grasshopper Effect Serves Pollutants onto Plates of Arctic Peoples," UN
Environment Programme, November 16, 2018, https://www.unep.org/news-and-stories/
story/grasshopper-effect-serves-pollutants-plates-arctic-peoples; Agamuthu Pariatamby
and Yang Ling Kee, "Persistent Organic Pollutants Management and Remediation,"
Procedia Environmental Sciences 31 (2016): 842–48.

106. Bruce Johansen, "The Inuit's Struggle with Dioxins and Other Organic Pollutants,"
American Indian Quarterly 26 (2002): 479–90.

107. Dusik, "Grasshopper Effect Serves Pollutants onto Plates of Arctic Peoples"; Arnold,
"Toxic Trouble as the Arctic Heats up."

108. Dusik, "Grasshopper Effect Serves Pollutants onto Plates of Arctic Peoples."

109. Doris Friedrick, "The Problems Won't Go Away: Persistent Organic Pollutants in the
Arctic," The Arctic Institute, July 1, 2016, https://www.thearcticinstitute.org/persistent-
organic-pollutants-pops-in-the-arctic/.

110. Watt-Cloutier, *The Right to Be Cold*, 137.

111. Watt-Cloutier, *The Right to Be Cold*, 141.

112. Brian D. Laird, Alexey B. Goncharov, and Hing Man Chan, "Body Burden of Metals

and Persistent Organic Pollutants among Inuit in the Canadian Arctic," *Environment International* 59 (2013): 33–40; Kavita Singh and Hing Man Chan, "Association of Blood Polychlorinated Biphenyls and Cholesterol Levels among Canadian Inuit," *Environmental Research* 160 (January 2018): 298–305; Friedrick, "The Problems Won't Go Away."

113. Sheila Watt-Cloutier qtd. in Johansen, "The Inuit's Struggle with Dioxins and Other Organic Pollutants," 484.

114. Frédéric Dallaire et al., "Effect of Prenatal Exposure to Polychlorinated Biphenyls on Incidence of Acute Respiratory Infections in Preschool Inuit Children," *Environmental Health Perspectives* 114, no. 8 (2006): 1301–5; Renée Dallaire et al., "Growth in Inuit Children Exposed to Polychlorinated Biphenyls and Lead during Fetal Development and Childhood," *Environmental Research* 134 (October 2014): 17–23; Laird, Goncharov, and Chan, "Body Burden of Metals and Persistent Organic Pollutants among Inuit in the Canadian Arctic"; Singh and Chan, "Association of Blood Polychlorinated Biphenyls and Cholesterol Levels among Canadian Inuit."

115. Singh and Chan, "Association of Blood Polychlorinated Biphenyls and Cholesterol Levels among Canadian Inuit."

116. Watt-Cloutier, *The Right to Be Cold*, 163.

117. Friedrick, "The Problems Won't Go Away."

118. "Persistent Organic Pollutants," Inuit Circumpolar Council, last modified July 10, 2022, https://www.inuitcircumpolar.com/icc-activities/environment-sustainable-development/persistent-organic-pollutants-pops/.

119. Thomson Reuters, "UN Agrees to Create Global Plastic Pollution Treaty," *CBC News*, March 2, 2022, https://www.cbc.ca/news/science/plastic-pollution-agreement-1.6369869#:~:text=The%20United%20Nations%20approved%20a,the%202015%20Paris%20climate%20accord.

120. Amanda Boyd and Chris Furgal, "Towards a Participatory Approach to Risk Communication: The Case of Contaminants and Inuit Health," *Journal of Risk Research* 25 (2022): 1–19.

121. Jody Wilson-Raybould, *"Indian" in the Cabinet* (Toronto: HarperCollins, 2021), 2.

122. Wilson-Raybould, *"Indian" in the Cabinet*, 187, 180.

123. Watt-Cloutier, *The Right to Be Cold*.

124. Government of Canada, "Pan-Canadian Framework on Clean Growth and Climate Change" (report, Ottawa, 2016), 1–78; Government of Canada, "A Healthy Environment and a Healthy Economy" (report, Gatineau, 2020), 1–79.

125. Eriel Tchekwie Deranger, "The Climate Emergency and the Colonial Response," Yellowhead Institute, July 2, 2021, https://yellowheadinstitute.org/2021/07/02/

climate-emergency-colonial-response/.

126. Little, "Iqaluit's Water Crisis Highlights Deeper Issues with Arctic Infrastructure."

127. Amanda Connolly, "Climate Change will 'Fundamentally' Shift How Canada, Allies Handle Arctic: NATO Chief," Global News, August 26, 2022, https://globalnews.ca/news/9087877/climate-change-canada-arctic-security-nato-trudeau-stoltenberg/.

128. Watt-Cloutier, *The Right to Be Cold*.

129. Torrealba, "From Inuit Nunangat to the Marsh."

130. Wilson-Raybould, *"Indian" in the Cabinet*.

131. Watt-Cloutier, *The Right to Be Cold*, xv.

132. Inuit Tapiriit Kanatami, "National Inuit Climate Change Strategy" (report, Ottawa, 2019), 1–44.

133. Alfred and Corntassel, "Being Indigenous," 610.

134. Wilson-Raybould, *"Indian" in the Cabinet*, 142.

135. "Canada Health Act," Government of Canada, last modified June 24, 2022, https://www.canada.ca/en/health-canada/services/health-care-system/canada-health-care-system-medicare/canada-health-act.html.

136. Josée G. Lavoie, Derek Kornelsen, and Yvonne Boyer, "Patchy and Southern Centric: Rewriting Health Policies for Northern and Indigenous Canadians," in Schiff and Møller, *Health and Health Care in Northern Canada*, 385.

137. Laureline Simon, "What Can Inuit Teach Us about Climate Change and Mental Health?," One Resilient Earth, April 28, 2021, https://oneresilientearth.org/what-can-inuit-teach-us-on-climate-change-and-mental-health/.

138. Eve Tuck, "Suspending Damage: A Letter to Communities," *Harvard Educational Review* 79 (2009): 409–27.

139. Inuit Tapiriit Kanatami, "National Inuit Strategy on Research."

140. Anang et al., "Building on Strengths in Naujaat."

141. Darrien Morton, "The Urban Indigenous Health Research Gathering" (report, Winnipeg, 2019), 4.

142. Inuit Tapiriit Kanatami, "National Inuit Strategy on Research," 6.

143. Liboiron, *Pollution Is Colonialism*.

144. Tuck, "Suspending Damage."

145. Wilson-Raybould, *"Indian" in the Cabinet*, 26.

146. Watt-Cloutier, *The Right to Be Cold*.

COVID-19 and Socioenvironmental Sustainability

Grassroots Strategies of Autonomy and Healing among and between the Ngigua in San Marcos Tlacoyalco, Puebla-Mexico

María Cristina Manzano-Munguía, Guillermo López Varela, and María Sol Tiverovsky Scheines

In Mexico and around the world, Indigenous communities faced COVID-19 in multiple and fragmented ways; local responses represented the first approach to a pandemic emergency among and beyond their communities.[1] Here we want to emphasize how COVID-19 triggered Ngigua grassroots constructions related to, but not limited to, sustainability and socioenvironmental balance with their practice of traditions.[2] Specifically, we look at how socioenvironmental practices revive the use of traditional medicine. Detailed information about rituals and the use of medicine plants is not addressed due to its sacred character. Out of respect, we only mention the types of activities performed during COVID-19 and the grassroots responses to the use, selection, and care of medicine plants. Thus, we look at the rain ritual and the sowing of rain while caring for the community *jagüey*, a natural water reservoir, where grassroots practices served to overcome the effects of COVID-19 among the Ngigua community (see figure 1). This water-collection method was widely used among land-based communities throughout Mexico. Periods with heavy rain were usually the high peak for their use. Today, this is still practiced and has the

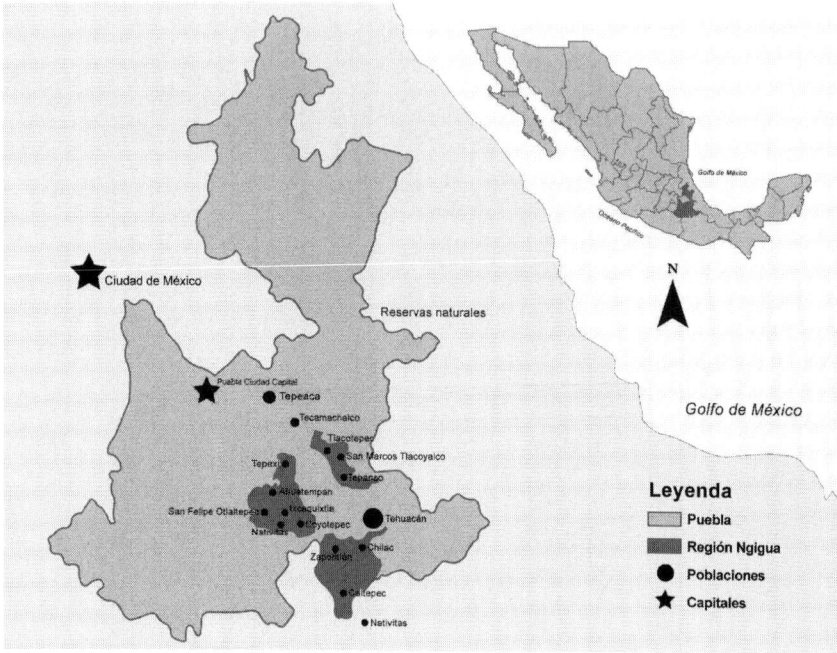

FIGURE 1. The Ngigua region. Google Maps (with modifications by López Varela).

same use but different meanings, which was worth exploring in our research. In other words, Indigenous populations implemented grassroots strategies that are seen as revitalization processes of their traditional ways to relate with diseases and their environment.

Manzano-Munguía and López Varela conducted fieldwork over two periods, in March and October of 2021. This included but was not limited to informal conversations, "deep hanging out,"[3] and interviewing key community members such as Elders, organic intellectuals (professors, students, and community leaders), and families. In the academic unit of San Marcos Tlacoyalco, the university population included 219 students, and the vast majority were women (approximately 72 percent), of which 18 percent were mothers (statistical information provided by the Intercultural University of the State of Puebla). Our participants (n=4) lived within the limits of the municipality of Tlacotepec of Benito Juárez including San Marcos Tlacoyalco and San José Buenavista. During our first season of fieldwork in March 2021 we were able to witness the contribution and relevance of educational performance in theater representations during the learning process and in language revitalization projects at the university level. Under the direction of Dr. López Varela, students (BA language and culture) undertook community liaison projects with enthusiasm and portrayed their traditions and their connections with Mother Earth through documentaries. We also annotated in our fieldwork diaries different cultural events (e.g., *Santujni*–Day of the Deaths, workshops, artistic performances), healing ceremonies, and academic activities at the Intercultural University of the State of Puebla. For example, the students at San Marcos documented the performance related to the water collection in the *jagüey*: *Tenchrua Ni Jinda ThueNcaa*.[4] Manzano-Munguía and community members produced a documentary titled *Kiense ja' an ná | ¿Quiénes Somos? | Who Are We?*[5]

In this study, the concept of organic intellectuals relates to Gramsci, and we recognize the leadership of Ngigua student intellectuals who achieve changes to the prevailing social order through their unique position of understanding their social conditions from below and acting as "functionaries" of the complex superstructures.[6] The organic intellectual is therefore forged from the bottom up—"from the structural base upwards"—and through political "participation in practical life, as constructor, organiser, 'permanent persuader' and not . . . [a] simple orator."[7] Moreover, we witnessed how the Ngiguas re-created the use of medicine plants, baskets of solidarity, and the use of the *jagüey* through the support and wisdom of their Elders.[8] Our fieldwork during the COVID-19 pandemic ignited, in a very provocative way, a rethinking and conceptualizing of grassroots constructions of care, support, revitalization, and endurance among community members in times of crisis.[9] To clarify, and in concurrence with Erik Wolf, our anthropological research began "with immersion in local experience and local knowledge," but still the "locals" remained the "other" in our research.[10]

It is important to note the composition of the student population at the Intercultural University of the State of Puebla: 72 percent self-identify as women, and 30 percent have already experienced motherhood. Their double role as intercultural students and as mothers precisely creates the bond for understanding the implications of motherhood and the sociocultural processes that may affect their territories and their care for nature. In this text, we wanted to underline the importance of the companions in the elaboration of a critical agenda of socioterritorial governance. Thus, our participants included two married women and the other two unmarried; however, all four of them

PARTICIPANTS FROM THE LANGUAGE AND CULTURE
AT THE INTERCULTURAL UNIVERSITY OF THE STATE OF PUEBLA

NAME	AGE	MAIN ACTIVITY	SOCIAL STATUS	PLACE OF ORIGIN
Pamela Díaz	20	Student and worker (mother)	Single	San Marcos Tlacoyalco
Renata Gómez	22	Student and worker (mother)	Married	San Marcos Tlacoyalco
Xóchitl Ramón	20	Student and worker (mother)	Single	San José Buenavista
Julieta Mayoral	23	Student and worker (mother)	Married	San Marcos Tlacoyalco

Information provided by Guillermo López Varela (2022).

experienced motherhood. We interviewed four female students currently pursuing their undergraduate degree in language and culture (see table), and most of them were involved in community projects as well as maternity. As noted, two students are single and the other two are married, and most of them are currently living in San Marcos Tlacoyalco with only one student from San José Buenavista. Our sample is significant given the patterns we were able to find across the experiences of our four participants in understanding healing and their bonds with their territories and bodies.[11] Moreover, the greater presence of women in our ethnographic research also corresponds to the fact that our region is characterized by medium migration rates. One in four Ngigua households migrate to the United States, and all households have at least one family member who migrates internally to the cities of Puebla, Tijuana, Los Cabos, La Paz, Nuevo Vallarta, Cancún, Playa del Carmen, and Mexicali. The men in this community are construction workers who tend to establish dynamic migratory circuits. Similarly, studying in the Ngigua context is considered a waste of time because gender roles for men see them as providers; so studying implies the impossibility of working and earning the necessary income to support their households. Therefore, women make up most of the student population. In addition, since the program of study is language and culture, there is a belief in the community that the graduates will be teachers within the community; from the perspective of the world of care, it is women who exercise these roles of educating the new generations. Historically, men have not yet been involved in the work of community care. It is women, from the multiple spaces they live, who undertake this journey from the awareness of their body they inhabit, a body indissolubly linked to the territory.

Here we want to note the following pertaining to the use of critical interculturality. In our reflection, critical interculturality is a transversal axis to fight inequalities in all aspects of daily life. Our experience accompanying the processes of defense of the territory, culture, language, and life of the Ngigua culture has helped us to understand different ways of strengthening the mechanisms of participatory and inclusive deliberation in their territories and learning traditional knowledge. There is a diversity of horizons in resistance that strengthen the linguistic and cultural autonomy of Indigenous people, making visible the failure of the cultural hegemony of national states in the praxis. At the heart of Ngigua cultural autonomy, nature is not a resource, object, or input that exists to be exploited. On the contrary, for the Ngiguas of San Marcos Tlacoyalco, in San José Buenavista, and any part of the Ngigua territory, a good living with all that surrounds it implies the collective joy of coinhabiting in synchrony and symbiotically with all that exists.

Consequently, we privileged the local experience and knowledge in understanding environmental sustainability from below rather than from above.[12] Bellfy noted how a "respectful relationship, developed over the millennia," referring to Indigenous people of the Great Lakes between the U.S. and Canadian territories: "their relationship with the natural world . . . understanding that if human beings take care of the environment, the environment will take care of them. . . . The entire relationship can be summarized as harmony and balance, based on respect."[13] As David McNab accurately puts it, "Indigenous knowledge comes from the Land. Indigenous people view their Lands and Waters, indeed all of Nature as the conjoining of Mindscape and Landscape."[14]

However, the environment has been disrupted historically from its balance due to capital interests, the exploitation of natural resources, humans as consumers, and how "Nature is being consumed."[15] In a crudely provocative way, Ward Churchill invites Indigenous and non-Indigenous readers, activists, government officials, academics, students, and the public in general to realize that "humanity is killing the natural world and thus itself."[16] "Postindustrialism," as Churchill refers to it, is the cause of the destruction of the environment and the natural order.[17] What Bellfy refers to as the "technological people" have destroyed the environment and continue to do so in a brutal and extended manner.[18] N. Scott Momaday calls for restoring traditional practices to gain equilibrium between humans and nature.[19] What seems to have been lost over the years is respect for the animal (buffalo) used for medicinal purposes; in the end, it was almost extinguished by the colonizers' exceeded hunting practices. In other words, Momaday remembered his grandmother told him of attending the last Sun Dance ceremony held annually among the Kiowa (north and west of the Wichita Range in Oklahoma). It was closely intertwined with "the restoration of her people in the presence of Tai-me . . . [the latter refers to] a vision born of suffering and despair. Take me with you Tai-me said . . . the sacred Sun Dance doll symbol of their worship."[20] The Kiowa Sun Dance was closely related to the buffalo hunt, their medicine tree, and their warrior spirit. As Momaday expressed it, "the last Kiowa Sun Dance was held in 1887 . . . [but] the buffalo were gone. In order to consummate the ancient sacrifice—to impale the head of a buffalo bull upon the medicine tree—a group of old Kiowa men bartered for a buffalo in Texas."[21] This was the last Sun Dance ceremony carried out. Here we want to emphasize the revitalization processes of traditional medicine in times of need and disease and the embedded practices among the Kiowas.

As Alexander Lesser, a cultural anthropologist, accurately noted: "the economic stability and security of the Indian tribes vanished" with the disappearance of the buffalo herds. Instead "want and hunger" prevailed.[22] Within this scenario, the revival of the Ghost Dance encouraged the belief of the return of "the buffalo and old Indian ways," the ruin of the white man, the fulfillment of the Christian precepts of living in brotherhood rather than the experiences of intertribal warfare, and their reunion with their deceased ancestors.[23] One of Lesser's major contributions was to demonstrate how Indian identities survived and revived despite systematic attempts to assimilate them into the European scheme. For instance, the Pawnees (Plains Indians) survived "after two hundred years of genocide or forced assimilation, [and] stubbornly persist[ed] in maintaining [their] American Indian identity."[24] Relatedly, the environment is closely entwined with the social, political, and economic contexts historically driven by and closely related to Indigenous constructions and understandings of the environment. McGregor noted "traditional ecological knowledge" where the major interest, according to Aimée Cree Dunn, was not only the "revitaliz[ation of] the old ways, . . . but it also involve[d] learning the values for protecting the land and all our relations in the present and future."[25] Here we explore the revitalization process of two traditional practices due to COVID-19, but before we launch into further details, we need to look at how Ngiguas experienced the pandemic.

Living the Pandemic

In 2022, Mexico's National Council of Evaluation for Social Development Policy (CONEVAL) reported that 46.8 million people lived in poverty, with four states accounting for the highest levels: Chiapas, Guerrero, Oaxaca, and Puebla.[26] The latter included 54 percent or 3,626,900 people living in poverty.[27] The CONEVAL measured the following poverty markers across the country: health services, education, housing, electricity, sewer and water services, social security, and safety. During the pandemic, it was relevant to access health-care facilities as well as medications, as earlier treatment saved lives. However, in many cases achieving this was not feasible. The pandemic made the challenges faced by Indigenous communities visible at the national level: most of them had to break the confinement and work (e.g., in the fields, service industries, construction, and as *jornaleros* [laborers], among others).

As mentioned in an earlier work,[28] the structural problems of the health system in Mexico (funds for personnel, medications, and access to health services) were clearly present through the lack of infrastructure needed in rural areas for COVID-19 patients and their health-care providers. The lack of ventilators and the need for intensive care units, medications, doctors, and health personnel illustrated the social and economic inequalities across the country. The state of Puebla represented one of the leading states with a higher number of Indigenous deaths (n=128) in December 2020.

Moreover, the lethality rate of COVID-19 among Indigenous people in the state of Puebla, as well as the critical vulnerability index within the municipalities of the Ngigua region, was around 32.5 percent.[29] In Puebla, 31 percent of deaths due to COVID-19 did not have any comorbidity, and 69 percent had at least one of them: hypertension and/or cardiovascular diseases (30 percent), diabetes mellitus (24 percent), and/or obesity (15 percent).[30] Adding complexity, the Ngigua region in Puebla is an area with a high, very high, and critical vulnerability index, where demographic, health, and socioeconomic dimensions converged to make its inhabitants vulnerable to and prone to die of COVID-19.[31] Consequently, we looked at the Ngigua community to illustrate the increasing challenges, necessities, and local responses endured by Indigenous communities during confinement.

From the Ground Up: Grassroots Constructions

The policy implemented by the government of Andrés Manuel López Obrador, president of Mexico, on March 22, 2020, halted all activities considered nonessential including, but not limited to, government offices, educational institutions, restaurants, entertainment centers, parks, and recreational activities, as well as hotels and tourist attractions. Moreover, students faced the challenge of distance education where poverty limited the use of and access to internet and electricity services. This in turn meant a complete failure to access educational programs. Rural areas represented one of the major challenges for accomplishing distance education. Consequently, the pandemic also revealed Indigenous students' educational challenges where "only 44.3 percent had a computer, 56.4 percent had access to internet, and 44.6 percent used a computer as an educational tool."[32]

Families living in San Marcos and other rural areas had to continue with their jobs and household endeavors. Quite often this meant breaching confinement. For instance, seasonal farm workers (*jornaleros*) continued working in Sonora and Baja California, those in the services areas in Los Cabos, Monterrey, Tijuana, and Playa del Carmen, and construction sector workers in the United States in North Carolina. Community members also reached out to family or friends who lived in the United States and Canada for money as their income abruptly changed. Thus, some of the students at the Intercultural University faced confinement as laborers, students, and mothers.[33] We now turn to examine Ngigua grassroots community initiatives, which represent the care, healing, and affection needed within the community during the first and second waves of the pandemic (spring 2020 and winter 2021). In other words, the production of roots and bonds were based on the values of use and not on a market economy as commodities;[34] the pedagogies were not subsumed to capital, patriarchal, and colonial relations in the sociocultural region of Ngiguas. In the following paragraphs, we want to demonstrate how grassroots strategies of autonomy developed into socioenvironmental sustainability projects where the use of traditional medicine expressed the care of bodies and thus environment sustainability practices unfolded.

Medicinal Plants and Environmental Sustainability

Earth is for us our mother, who gives life, who feeds and embraces us. We belong to her; that is why we are not the owners of any land. Between mother and children, the relationship is not related to property, but rather a sense of belonging. Our mother is sacred, and because of her, we are sacred too. The Earth as territory is ours. Each one of the elements of nature provides a specific function that is necessary for everything, and this concept of integration is present in each aspect of our life. It is neither possible to separate the atmosphere from the ground nor from the soil. It is the Earth a total space. It is from within this territory that we learn to give sense to equity. . . . Because the living species, it is like that because the Earth is life. For us, the Ngiguas, the Earth is not something that we can divide into small huts; rather it is something essential, with all its parts, with all that exists in nature, with all that produces within it, and in relation with its knowledge.[35]

The cultural topographies of the Ngigua body teach us that we cannot think about embodiment and the processes of subjectivity attached to them without understanding territorialities as symbolic appropriations of what we inhabit and what inhabits us. The corn seed has a word and a memory, the ant, the cloud, the morning dew, the colors of the sunset, and the silence that crosses the starry sky. There is nothing that exists in the Ngigua culture that is not told how it was born. What anthropologists, such as Alicia Barabas, have called ethnoterritories are ways of inhabiting the experience of existing without commodifying (market value) or objectifying the life that surrounds us.[36] Ngigua cultural, therapeutic, medical, and traditional health practices teach us that nature (e.g., medicine trails, backyard medicine plants), others dialogue, and the memories of our ancestors can never be commodified. Therefore, healing in the Ngigua context goes through the experience of protecting the plants that are the memory of the grandmothers who transmit their knowledge through dreams, weavings, and the memory that inhabits the words that are pronounced when accompanying those who suffer. This is unlike the urban experience, where the cure comes through the act of buying medicines and medical consultations, where there is never an awareness that if the plants cease to exist, everything will cease to exist.

In the community of San Marcos Tlacoyalco there were aesthetic-political and sociocultural resistances from the Ngigua culture that were strengthened, paradoxically, within syndemic contexts: the *thengijna chooni naa ko naa ni* (*mano vuelta*) or the *thi tenkininxini sincheeni* (*promesa*) as true pedagogies of the communitarian *apapacho*,[37] which in turn meant a criticism to extractivism in all its forms. *Mano vuelta* (hand in return) is a practice commonly known among Indigenous communities across Mexico. This practice includes, but is not limited to, working without receiving any payment. *Promesa* has a similar meaning to *mano vuelta* where a nonmonetary return is expected in exchange for work, a community activity, care, providing food or resources. An *apapacho* is a Nahuatlism used in different regions across Mexico. It refers to the process of being together and keeping closer. These activities fed a reflection situated within the paradigm of the Ngigua communality in diverse horizons, such as the use of medicinal plants while being in the *temazcal* (a site made of adobe utilized for taking a hot bath with medicinal plants, usually linked to healing purposes and the sacred, whose use is dated before colonization), the request for rain (e.g., the *jagüey*), the baskets of solidarity around the *guajolote* (turkey), and the dance or the care of the *jagüey* that inhabit the region. These pedagogies to a greater extent healed the bodies/territories of Ngiguas during the COVID-19 pandemic, producing the care of affections, of rootedness and bonds, attached to the values of territory and environment.

The pedagogies addressed relate to care and connect with their diverse forms of life within their community. Their vision is neither anthropocentric nor reductionist about their surroundings. When Ngiguas' wise women and healers undertake therapies to heal an illness, it usually relates to understanding what inhabits the bodies/territories of those who are sick and seeking attention. The healers emphasize that only by healing the territories it is possible to heal the bodies. From them we heard for the first time the enunciation of bodies/territories, since there is no distinction in Ngigua that separates these two forms of existence. Throughout the ethnographic fieldwork, we were able to understand that our bodies get sick because the territory is getting sick. For this reason, the community teachers emphasized the importance of the territory being inhabited not only by human beings but also by a wide range of life-forms that must be preserved and with whom we must learn to walk and talk in spirit and as living species. The ancestors, the nonhuman presences or owners of the mountain, the rain, and all that lives (the *chinentle*), the beings that are both animal and human (the *nahuales*), and the messengers of the forces that sustain life (the *xantilles*) are involved in the delicate balance that exists between the living and the dead.

FIGURE 2. The chain of hills in San Marcos Tlacoyalco. Gabriela Hoyos Cansino, 2021. Copyright permission to publish this photograph from the research project entitled "Multiculturalism and Multilingualism: Indigenous Return Students within the Intercultural Education in Mexico. Towards the inclusion of a multilingual and pluricultural education among Indigenous return migrants at the Intercultural University of the State of Puebla, San Marcos Tlacoyalco." Financed by the Vicerrectoría de Investigación y Estudios de Posgrado (VIEP-BUAP 2020–2023). Principal Researcher Dr. Manzano-Munguía.

The COVID-19 pandemic revived the grandparents' concerns to understand that the climate is changing and, as they told us, the earth also needs to rest and human beings need to do something to restore the balance that we have broken with everything that exists, lives, speaks, and walks. Therefore, in the Ngiguas' cultural horizons, the cure for a disease can never be thought of as an individual entity but always as a collective one. It can never work as a consequence of physiological factors alone but always includes psychosomatic aspects. As we mentioned, when people in the community get sick it is because something in the territory is also unbalanced, but they also care for the sick including the community who safekeeps and takes care of those in need. From a political ethic of collective care, everyone is aware that neither illness nor health is eternal and we always need others to survive.

Thus, the care for and use of medicine plants represent a revival process of taking back their roots, where the role of wise men and women in the millenary Ngigua culture is related, historically, to the ritual mediation established with the hills. The mountains and mountain ranges are the origin of culture and community life (see figure 2). The shamans performed rituals for the rainy season, as water is a scarce natural resource in this semidesertic environment. The community organizes themselves to clean the hills and ravines; they are also food providers for those who enter and clean the hills. As one Elder mentioned: "not everyone could enter the hill, only the right ones because not everyone has the gift to do the work that was done and most of them are men. For this reason, they stopped practicing the knowledge they had because there are no longer shamans with the same courage to do it."[38]

The work of rain pediment or cleaning of the hills and mountains is performed by men due to the belief that several vital processes prevent women from performing this type of work. For example, a woman who is breastfeeding or pregnant or has her menstrual period cannot perform this type of activity. Nor when they have performed work that implies that their bodies have "warmed up," such as making tortillas or using medicinal therapeutics. There is a sexual division of labor that prevents women from performing this particular work. It is considered there are heavy days such as Tuesdays or Fridays when these tours cannot be done because the owner of the mountain can lose people. Likewise, asymmetrical relations have been perpetuated in their community so that only men can do these activities, which invites us to think about Ngiguas' sociocultural practices under antipatriarchal keys and also to question the wise men and women about the implications of this sexual division of community work and bonding projects.

Located at the heart of a protected reservoir, community members, Elders, and students conducted community-bonding projects connected with their culture, environment, and relations (see figure3). As *Mzinegiizhigo-kwe* Bédard stated for Indigenous people: "Revitalizing the old ways by learning the traditional knowledge of our Native . . . ancestors and adapting it to today."[39]

In terms of acute economic and health vulnerability, such as COVID-19, the recovery of Ngigua biocultural landscapes for the strengthening of community health and well-being is currently being deployed. In the following paragraphs, we

FIGURE 3. Trails of medicine plants in San Marcos Tlacoyalco. Iván Romero Serrano. Copyright permission to publish this photograph from the research project entitled "Multiculturalism and Multilingualism: Indigenous Return Students within the Intercultural Education in Mexico. Towards the inclusion of a multilingual and pluricultural education among Indigenous return migrants at the Intercultural University of the State of Puebla, San Marcos Tlacoyalco." Financed by the Vicerrectoría de Investigación y Estudios de Posgrado (VIEP-BUAP 2020–2023). Principal Researcher Dr. Manzano-Munguía.

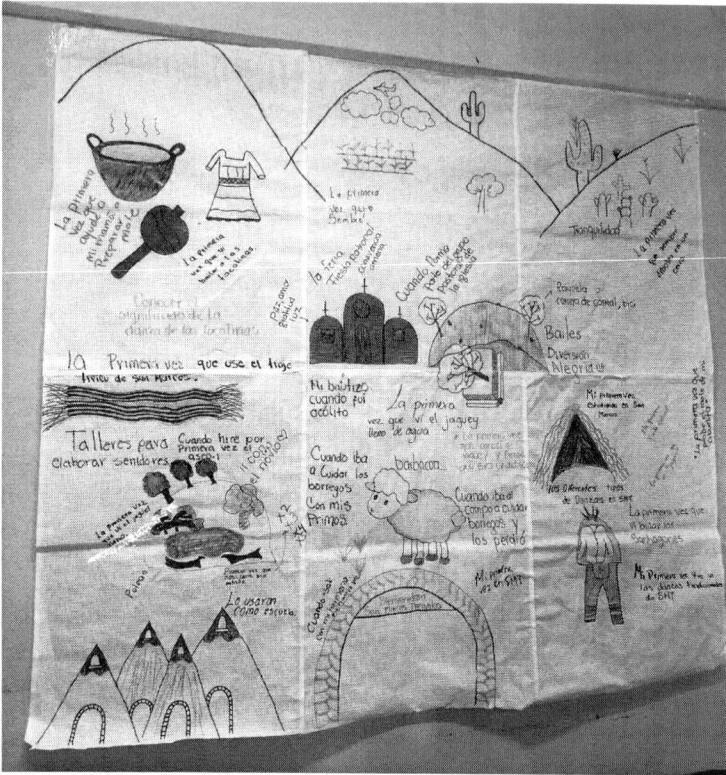

FIGURE 4. Social, emotional, and experiential mapping of San Marcos Tlacoyalco.
Children and youth of San Marcos Tlacoyalco and San José Buenavista, 2021–22. Copyright written permission to publish this photograph was granted to Dr. Guillermo López Varela, research professor at the Intercultural University of the State of Puebla, campus San Marcos Tlacoyalco; he is also a community liaison.

illustrate some of our findings through the work of students at the Intercultural University of the State of Puebla campus San Marcos who are currently undertaking different community projects that have a positive impact on the health and well-being of community members, and also provided written consent to publish their photos and to include the data collected during their interviews.

One example is the Ngigua intercultural community recipe book that connects the defense of their territory and social cartographies that document various organizational processes of their ethnoterritories in Puebla. Recently a community herbarium illustrated their medicinal, edible, ritual, and therapeutic plants with a focus on scientific dissemination. It also included universal

FIGURE 5. Botanical collage, "Bio-cultural Ngiguas Landscapes Workshop." Children and youth of San Marcos Tlacoyalco and San José Buenavista, 2021–22. Copyright written permission to publish this photograph was granted to Dr. Guillermo López Varela, research professor at the Intercultural University of the State of Puebla, campus San Marcos Tlacoyalco; he is also a community liaison.

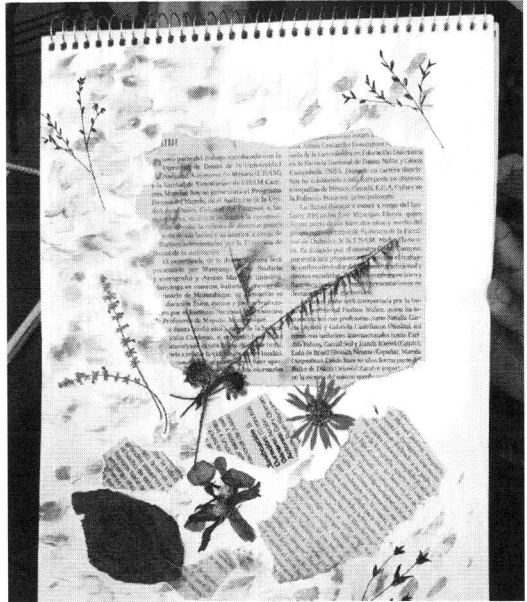

FIGURE 6. Fragment of a community herbarium. "Bio-cultural Ngiguas Landscapes Workshop." Children and youth of San Marcos Tlacoyalco and San José Buenavista, 2021–22. Copyright written permission to publish this photograph was granted to Dr. Guillermo López Varela, research professor at the Intercultural University of the State of Puebla, campus San Marcos Tlacoyalco; he is also a community liaison.

access to knowledge for Ngigua children and youth and the encouragement of early scientific vocations (see figure 4). Here the emphasis is on their grandparents and wise men who threshold the care of Ngigua embodiments and territorialities. The ethnoterritories and the ritual practices performed implied the balance between healthy bodies and the care of the territory through socioenvironmental practices such as the cleaning of the hills and the rituals for waterfalls.[40] Specifically, during the peak periods of the pandemic, the *curanderas* (healers) were continually sought to perform a ritual and provided medicine plants to the individual who lost his or her balance (health) and suffered from *frialdad* or a cold (see figure 5).

The *curanderas* hold the knowledge for healing and restoring the balance in one's health. For instance, the *quebrantamiento* (feeling broken) is when you receive *un aire* (a blow) and "the tendons of the body move or perhaps a bone moves or you shake hard from a blow."[41] The *curanderas* have the gift of healing. They know what type of medicine plants to use to heal the body of the person who is suffering from *rompimiento* (brokenness). Growing with and among nature included knowing your territory and its surroundings, Elders, shamans, and *curanderas*, as well as organic intellectuals, and knowing the type of plants available for healing and health restoration. Mother Earth is the provider of the gift of life (see figure 6). The *curanderas* "heal children and adults who suffer from sickness; there are two types of sickness, when they turn purple or white. That is when they use all kinds of [description of the plants used] . . . and water . . . composed also of spirit water, then they put it on the body of the sick person; . . . they call their spirits when they are scared."[42] The figure about the community herbarium (see figure 6) serves as an illustration of the type of medicine plants that are available throughout the healing paths trails (*senderos medicinales*). Some medicine was used to cure and restore health after COVID-19, and in other instances, the *curanderas* performed their rituals, including but not limited to the use of the *temazcal* to restore their health and balance. The use of water while being inside the *temazcal* was closely related to the request of rain and the use of the *jagüey*.

The Request for Rain and the Preparation of the *Jagüey*

The challenges the Ngigua face in the semidesert area of Puebla due to water stress in the Valley of *Tehuac*án led their ancestors to carry out rituals related to the care of the Ngigua ethnoterritories and embodiments.[43] One example is worth noting: the *jagüey* as life provider within the communities (see figure 7).

To understand these rituals, it is relevant to look at their collective scale, the scenario of the new community pedagogies implemented, and their historical resistance to the multiple energy extractivist projects deployed at San Marcos. This is exemplified by the student brigade of the *Teatro histórico Ngigua, Xra Juinche'e xan* (Ngigua Historical Theater) where their performance depicted the consequences and dimensions of extractivism and the importance of their translation into political aesthetic language.

Every January 1, as the work of the Ngigua Historical Theater Community Company demonstrates, in the community of San Marcos Tlacoyalco a ritual is performed that includes the burying (sowing) in the *jagüey* at the center of

FIGURE 7. The community *jagüey*, San Marcos Tlacoyalco. Iván Romero Serrano. Copyright permission to publish this photograph from the research project entitled "Multiculturalism and Multilingualism: Indigenous Return Students within the Intercultural Education in Mexico. Towards the inclusion of a multilingual and pluricultural education among Indigenous return migrants at the Intercultural University of the State of Puebla, San Marcos Tlacoyalco." Financed by the Vicerrectoría de Investigación y Estudios de Posgrado (VIEP-BUAP 2020–2023). Principal Researcher Dr. Manzano-Munguía.

the community a jar with sacred water. The small jar must be accompanied by white candles decorated with ribbons and white flowers. It is also customary to bring water from the *Aljojuca* Lagoon in Puebla and the *La Mixteca* Lagoon in Oaxaca.

All these jars are buried in the middle of the *jagüey* and at the foot of it, with the belief that it will conserve water throughout the year and will ignite community life. In addition to the ritual work in the community's navel *jagüey*, the ravine from which the water flows directly from the mountain must be made ready. When asked how the people from San Marcos Tlacoyalco prepared the ravine, Elder don Miguel Sánchez Damián noted:

> It consists of cutting mesquite branches and pitchforks to bury both sides of the ravine and then to put the branches between the pitchforks so that when the rain begins the water pours into the *jagüey* without garbage, stones, branches or any other plants that it drags from its path. For the community of San Marcos Tlacoyalco it was their most effective water filter system so that it [the water] arrives clean to the *jagüey*. This type of filter was placed at various points in the ravine, and the last one was left at the gate of the *jagüey* for the filtration and purification of water. This work was done in the month of February.[44]

Subsequently, on the elaboration of community filters used during the Ngigua agricultural cycle, Ramón and Mayoral explained:

> In the community of San Marcos Tlacoyalco during the month of March, the preparation of the *jagüey* begins. When they finished placing the filters in the ravine, they proceed to do the work in the *jagüey*. It consists of performing tasks to measure; they used a stick to measure the width, length, and depth of the task. They [also] used a shovel, a pick to dig, and to carry the soil they used a basket, which they made themselves.[45]

When the tasks described were carried out, the Ngiguas elaborated a "flower ritual" that consists of the following:

> In the first rain the people wait in the neighborhood for the column at the edge; . . . from there they waited until the *jagüey* received the water that came from the hill, they threw . . . [different plants] in order to . . . , they also threw . . . as a symbol of

joy and also to warn the people who are in the church to . . . as a symbol of thanks and joy for the rain. This ritual is performed after the first rain.[46]

If the rain has been very scarce, they are delayed, and especially before the agricultural cycle begins in the so-called request for rain to the deities of the hill a shaman carries out this ritual. This activity begins on December 9 with the cleaning of the *ruakanchrje* hill Yolotepec and ends on December 24 at the *Cerro Colorado* (Red Hill) or *jna jatse*. The offerings to the hill deities consist of different natural items. The shaman carries items that are used to illuminate the caves where the Elders (male) go to clean, approximately three hundred meters deep; they carry spoons to mark their way back. These spoons serve as a return guide from the unknown to the known and from the known to the unknown. Around December 24, a great celebration is held at *Cerro Colorado*:

> They take casseroles with *mole, barbacoa, tamales*, but all this food is prepared without salt, especially for the deities of the hill; the male is called *chanincle*, female is called *chanincla* and the porter. It is decorated, as if it was a party; they coexist among the deities of the hill; also the people who go coexist among them outside the cave with music and food. The people leave the place; after a few days they return to pick up the dishes they brought the food [in]; they pick up the dishes, they bring them back, and those dishes are kept for the following year.[47]

If after this ritual it does not rain, as in the present year when they are narrating their experiences, they visit the hill again to perform the same ritual. It should also be performed on June 15 so that the crops are harvested or later with the danger of snow. Finally, the healer and his companions deliver the report of the work done during the rain petition to the hill. The shaman and his companions arrive at the community; the population goes out to wait, with food to share, and celebrate the work performed. They arrive at the presidency of the auxiliary board, and there on the floor another ritual is performed and other people leave those items at the hills of San Marcos Tlacoyalco to bring rain to their town.

On the use of traditional medicine in the context of rituals of healing as a Ngigua ethnoterritorial ritual, Renata Gómez stated:

> Finally, the [items] that are removed when they cleaned the caves, the auxiliary president is in charge of [disposing of the items] in the community [location]. This

ritual has been performed every year since pre-Hispanic times. It stopped being practiced approximately thirty to fifty years ago. That is why the rain feeding the land is no longer sown, as it only depends on rainwater.[48]

The water heals; this is an important element for creating a balance between the community and the environment. This natural resource connects in multiple and complementing manners among and between Ngiguas. During COVID-19, specific rituals such as the one described briefly took place, but none of them called upon health restoration through water use (excessive washing of hands and clothes plus other public or common areas being disinfected). Our work represents just an example of the complexity lived from March 2020 until restrictions were lifted during the spring of 2021.

Conclusion

As we have previously reported, the production of practices that intensify the care of human bonds with other forms of life is significant and in need of an explanatory matrix of the daily life of the Ngigua people living in the region of Puebla. Through the evocation of rituality contextualized in a festive agricultural cycle that embodies practices linked to rain rituals and medicine plants, we accounted for the importance given by the Ngigua communities to the care provided in relation to the environment, the revival of their traditional knowledge practices, and the restoration of health within their bodies. Grassroots Ngigua constructions on the uses and effects of medicinal plants, rituals, shamans, and *curanderas* for their significant labor during the peaks of the pandemic (COVID-19) illustrate what ritualized environmental knowledge means while asserting the use of local resources for healing during this medical emergency. In other words, the community of San Marcos Tlacoyalco revived their grassroots healing knowledge and practices based on their use of medicine plants and healer practice in the face of COVID-19. Their stories of healing and autonomy are worth noting and are a path to learning. Here nature is being consumed for healing purposes and the restoration of health in times of COVID. Further research is needed to include the perspectives of the biomedical personnel and their healing paths during the COVID-19 pandemic at San Marcos. Also, studies need to reflect community initiatives based on traditional healing practices versus biomedical strategies for health restoration and healing practices at the local and national level.

NOTES

We acknowledge the funds received by the *Vicerrectoría de Investigación y Estudios de Posgrado (VIEP)* at the *Benemérita Universidad Autónoma* de Puebla since 2021 for carrying out fieldwork and our research at San Marcos Tlacoyalco, Tlacotepec de Benito Juárez. We express our gratitude and a big thank you to the Ngigua community for their hospitality in 2021. A special thank you to the Intercultural University of the State of Puebla in San Marcos Tlacoyalco for their help and support , and Dr. Tatiana Konrad for her diligent work as editor. Nevertheless, we are responsible for the information and analysis provided in this essay.

1. See for instance the "Indigenous People and Pandemics Conference" (May 15–16, 2020) and "Pre-Meeting Workshop" (May 13) organized by the Centre for Advanced Study at the Norwegian Academy of Sciences and Letters in Oslo, Norway (2023) where Indigenous people presented about their local responses to COVID-19, and some were not related to government assistance or support; for details, see https://cas-nor.no/events/indigenous-peoples-pandemics-conference.

2. Guillermo López Varela and María Cristina Manzano-Munguía, "Governance, De-commoditization and Communality: The *Ngiguas* of San Marcos Tlacoyalco in Puebla, Mexico," *Ethnologies: Journal of the Folklore Studies Association of Canada* 43 (2021): 145–64; María Cristina Manzano-Munguía and Guillermo López Varela, "The Ngigua Community Jagüey: A Sanctuary of Indigenous Resistance Today," *The New Polis: A Journal of Critical Theory, Social Analysis and Political Philosophy and Theology* 1 (2022): 130–43, http://journal.thenewpolis.com/archives/1.1/index.html.

3. Renato Rosaldo, *Culture and Truth: The Remaking of Social Analysis* (Boston: Beacon, 1989).

4. For instance, "Vinculación comunitaria," UIEP Oficial, Facebook, September 25, 2022, https://www.facebook.com/watch/?v=767775437658840, and Teatro Histórico Ngigua (untitled video), YouTube, June 3, 2021, https://www.youtube.com/watch?v=6R8c-JlRVJ0&t=3s).

5. "Kiense ja' an ná | ¿Quiénes somos? | Who Are We?," Cultura Documental, YouTube, April 29, 2022, https://www.youtube.com/watch?app=desktop&v=Lax9y4zOKZM.

6. Antonio Gramsci, *Selections from the Prison Notebooks*, ed. and transcribed by Quintin Hoare and Geoffrey Nowell Smith (New York: International Publishers, 1971), 10, 418.

7. See Gramsci, *Selections*, 10.

8. For details about food baskets or baskets of solidarity, see López Varela and Manzano-Munguía, "Governance, De-commoditization and Communality."

9. For the adequacy of the term "crisis," see Erik Wolf with Sydel Silverman, *Pathways of Power: Building an Anthropology of the Modern World* (Berkeley: University of California Press, 2001).

10. Wolf, *Pathways of Power*, 51.

11. For community projects, see María Cristina Manzano-Munguía, Iliana Viridiana Roa González, and Iván Romero Serrano, "Xra Nche'e Ni Kunixin Rajna: Prácticas de Vinculación Comunitaria en la Universidad Intercultural del Estado de Puebla, San Marcos Tlacoyalco," *Mirada Antropológica* 18, no. 24 (2023): 58–77.

12. Wolf, *Pathways of Power*, 51.

13. Phil Bellfy, "Introduction," in *Honor the Earth: Indigenous Responses to Environmental Degradation in the Great Lakes*, ed. Phil Bellfy (East Lansing: Ziibi Press, 2014), xvii.

14. David McNab, "Weather or Not? Weather and the Environment in the Indigenous Thought and the Written Word of Ezhaaswe (William A. Elias—c. 1848–1929)," in Bellfy, *Honor the Earth*, 91–105, 91.

15. Ward Churchill, "Decolonization: A Key to the Survival of Native North America," in Bellfy, *Honor the Earth*, 11.

16. Churchill, "Decolonization," 4.

17. Churchill, "Decolonization," 12.

18. Bellfy, "Introduction," xviii.

19. N. Scott Momaday, "From *The Way to Rainy Mountains*," in *Constructing Nature: Readings from the American Experience*, ed. Richard Jenseth and Edward E. Lotto (Upper Saddle River, NJ: Prentice Hall, 1996), 229–50.

20. Momaday, "From *The Way to Rainy Mountains*," 233–34, 230, 232.

21. Momaday, "From *The Way to Rainy Mountains*," 234.

22. Alexander Lesser, *History, Evolution and the Concept of Culture: Selected Papers by Alexander Lesser* (Cambridge: Cambridge University Press, 1985), 119.

23. Lesser, *History*, 85, 119.

24. Lesser, *History*, 146.

25. Deborah McGregor, "First Nations, Traditional Ecological Knowledge and the State of the Great Lakes Ecosystem," in Bellfy, *Honor the Earth Indigenous Responses to Environmental Degradation in the Great Lakes*, ed. Phil Bellfy (Michigan: Ziibi Press, 2014), 109–30, 111.

Aimée Cree Dunn, "Listening to the Trees: Traditional Knowledge and Industrial Society in the American Northwoods," in Bellfy, *Honor the Earth*, 131–64, 154.

26. "Medición de la pobreza: Probreza por Localidad Urbana 2020," Consejo Nacional de Evaluación de la Política de Desarrollo Social (CONEVAL), https://www.coneval.org.mx/Medicion/Paginas/pobreza_localidad_urbana.aspx.

27. "Medición de la pobreza," CONEVAL.

28. Jorge Gómez Izquierdo, Sol Tiverovsky Scheines, and María Cristina Manzano-Munguía, "Pandemias y Modelos de Control Social: Reflexiones Foucaultianas Contemporáneas," in *Elementos 121* (Puebla: Benemérita Universidad Autónoma de Puebla, 2021), 71–76.

29. Dirección de Información Epidemiológica, "Décimo tercero informe epidemiológico 2021 de COVID-19 en la población que se reconoce como indígena," Gobierno de México, https://www.gob.mx/cms/uploads/attachment/file/660822/COVID-19_poblacion_indigena_2021.07.08.pdf.

30. Dirección de Información Epidemiológica, "Décimo tercer informe epidemiológico 2021."

31. Manuel Suárez Lastra et al., "Índice de vulnerabilidad ante el COVID-19 en México," *Investigaciones Geográficas* 104 (2021), https://doi.org/10.14350/rig.60140.

32. Teresa Moreno, "Con computadora, solo el 5% de los estudiantes pobres en el regreso a clases," *El Universal*, August 24, 2020, https://www.eluniversal.com.mx/nacion/con-computadora-solo-5-de-los-estudiantes-pobres-en-el-regreso-clases; María Ibarra, "La Universidad Autónoma Indígena de México: experiencias y retos de la educación virtual," in *1er Foro La educación virtual y las Universidades Interculturales en tiempos de pandemia* (Universidad Intercultural de Chiapas, 2020), on-line meeting for intercultural universities.

33. See López Varela and Manzano-Munguía, "Governance, De-commoditization and Communality."

34. Commodity is based on profit and hegemony in the market economy.

35. Informal interview with Lupita Flores (Elder), October 2021.

36. Alicia Barabas, "Etnoterritorialidad sagrada en Oaxaca," in *Diálogos con el territorio: Simbolizadones sobre el espacio en las culturas indígenas de México*, ed. Alicia Barabas, vol. 1 (Ciudad de México: Conaculta-INAH, 2003), 145–68.

37. Following Montemayor, *apapacho* means hugging with affection, showing care and feeling beloved by someone, including affection: Carlos Montemayor, ed., *Diccionario del Náhuatl en el Español de México* (México D.F.: Universidad Nacional Autónoma de México–Gobierno del Distrito Federal, 2007); Guillermo López Varela, "Pedagogía del apapacho entre los Ngiguas de San Marcos Tlacoyalco; contra-pedagogías dialógicas no estadocéntricas en contextos sindémicos," *XIV Jornadas de Sociología. Facultad de Ciencias Sociales* (Buenos Aires: Universidad de Buenos Aires, 2021), https://cdsa.aacademica.

org/000-074/629.pdf.

38. Interview with Pamela Díaz, Renata Gómez, Xóchitl Ramón, and Julieta Mayoral, 2022. Interviews conducted by students of the undergraduate course related to language and culture.

39. Dunn, "Listening to the Trees," 154.

40. Sabino Martínez Juárez, "Territorialidad Sagrada en San Marcos Tlacoyalco, Aproximaciones Teóricas y Etnográficas," in *Territorios Indígenas, Educación e Interculturalidad en la Región Sureste de Puebla y Sur de Veracruz*, ed. Felipe Galán, Sabino Martínez, Guillermo López Varela, and Alejandra Gámez (México: Benemérita Universidad Autónoma de Puebla, 2022), 79–85.

41. Interview with Xóchitl Ramón and Julieta Mayoral, March 2021.

42. Interview with Ramón and Mayoral, March 2021.

43. López Varela and Manzano-Munguía, "Governance, De-commoditization and Communality."

44. Interview with Miguel Sánchez Damián, March 2021.

45. Interview with Ramón and Mayoral, March 2021.

46. Interview with Julieta Mayoral, March 2021.

47. Interview with Mayoral, March 2021.

48. Interview with Renata Gómez, March 2021.

The Stop Cop City Movement

Environmental Racism, Community Policing, and Necropolitics in Post–COVID-19 Atlanta

Juan M. Floyd-Thomas

t took fifty-seven gunshots to fatally wound Manuel Estaban "Tortuguita" Páez Terán and jumpstart an ad hoc albeit growing environmental justice and abolitionist movement. The killing of Páez Terán, a queer, Indigenous-Venezuelan environmental activist, by Atlanta police officers on January 18, 2023, has been viewed by many supporters as brutal retaliation for their activism to protect a heavily forested area in Atlanta's southwestern section.[1] According to *The Guardian*, Páez Terán's murder sadly became the first time in modern U.S. history an environmental activist has been shot and killed by police officers during a protest.[2] Demonstrations and vigils were held in several cities in the United States and internationally in reaction to the shooting. Many organizations have condemned the killing, and it has even prompted some members of Congress to call for an independent investigation of the events surrounding Páez Terán's death. Meanwhile, there has been a months-long siege by forest-dwelling protesters to prevent the construction of this facility derisively called "Cop City" by its opponents as well as stop the imminent devastation to the woodlands surrounding the proposed training center.

Several months prior to this fatal incident, protesters had been engaged in a prolonged conflict over a massive police training facility that the city planned to build in this area. The Atlanta Police Foundation and the municipal government proposed the project (its official name being the Atlanta Public Safety Training Center), a $90 million police training facility, slated to be erected on an eighty-five-acre campus located on the outskirts of the city proper. Notably, the Cop City campus is designed to serve as a state-of-the-art mock-urban space replete with homes, convenience stores, and even a faux nightclub in the heart of a forest. As a result, this development scheme has drawn the ire of an eclectic combination of nearby residents, police reform activists, environmentalists, and even advocates for the homeless in an urban area desperately in need of greater infrastructure and social services.

In many regards, the Stop Cop City movement is a sort of synthesis of recent social movements in the United States and elsewhere. Like Occupy Wall Street over a decade ago, Stop Cop City features a strategically localized encampment against the neoliberal nexus of state and corporate power hellbent on imposing austerity on incredibly underserved and disenfranchised citizens of color. Akin to the NoDAPL movement at Standing Rock, the standing public gathering of protesters takes the form of an infrastructural barricade against the unrelenting expansion of statist and corporatist power. It also operates as an unabashedly outspoken climate defense of the Weelaunee Forest—once home to the Muscogee Creek peoples—in a metropolis with the greatest density of tree canopy of any major urban area in the United States. Most importantly, like the Black Lives Matter (BLM) uprisings circa the summer of 2020, this movement confronts the uniquely racialized history and inherently racist scope of policing in the United States as the most immediately deadly manifestation of the carceral state. An amalgamation of recent history only partly explains the almost incredible intensity of the backlash the organizers and protesters have faced. Although their violent interactions with the Atlanta Police Department have been quite alarming, it is the escalating levels of overreach by city and state government officials—especially acts of opaque decision-making, illogical policies, abusive practices, vindictive charges, and malicious prosecution—enacted with the clear intention to quash potential dissent that, in turn, could result in ruined landscape and ravaged lives.[3]

While the metropolitan government accepted the proposal to build Cop City, this mobilization was occurring at a particularly fraught moment in the city's recent history. On the one hand, the Cop City proposal emerged in city hall while perceptions of urban crime were mounting in 2021. The movement against Cop City in Atlanta began simultaneously as a campaign to derail efforts by the metropolitan and state governments toward overpolicing in response to a perceived spike in urban crime in the wake of the COVID-19 pandemic while it is also a struggle to defend the Weelaunee Forest against decimation. Moreover, while much of the city is still reeling from the impact of huge BLM protests in the summer of 2020 sparked by the murders of George Floyd, Rayshard Brooks, and others at the hands of law enforcement agents, the land designated for this new police outpost was originally named by the Muscogee Creek, an Indigenous population forcibly displaced by white settlers and the U.S. federal government from the land in the early 1800s during the infamous Trail of Tears. Some years later this area became the site of the notorious Old Atlanta Prison Farm on 1,248 acres in southeast Dekalb.[4] During a "Week of Action" in early 2023 meant to bring the construction of Cop City to a halt, a disturbance erupted out of a planned demonstration among hundreds of activists. The protestors engaged in property damage of construction infrastructure around a security outpost set up by police. Police responded, rushing to Weelaunee's People Park, where many participants in the Week of Action were attending a music festival in honor of Tortuguita. The fight came to a head when hundreds of activists breached the construction site, burned police cars and construction vehicles, and damaged property at the site. Since that confrontation and subsequent encounters with local and state law enforcement, this has resulted at the time of this writing with the arrest of more than fifty protesters, with more than half of these individuals being charged with committing acts of domestic terrorism.

Focusing on Atlanta's history over the last fifty years, this essay looks closely at the continued patterns of residential segregation, many of which exist in current legislation and regulatory policies despite the common beliefs that overtly racist policies had been dismantled as vestiges of the past. Drawing on social science research, policy analysis, and archival materials, the essay reveals the *longue durée* of the impact Atlanta's racial hypersegregation has had on its populace of color, from urban deforestation to toxic pollution to police brutality. Beginning with an analysis of this current political moment, the essay seeks to move beyond simply describing and decrying urban problems. Lastly, this is an exploration of how and why this grassroots struggle to preserve a forest on the outskirts of Atlanta became an unlikely new battleground in debates over neoliberal policies, residential hypersegregation, environmental racism, community control, and criminal justice reform is the main thesis of this essay.

From Plantations to Playground Cities: White Flight, the Right to Comfort, and the Paradox of Black Atlanta in the Contemporary Neoliberal Imagination

Although Atlanta has undergone numerous challenges and changes in many ways as it has transformed itself to meet the shifting reality of the contemporary world to become the so-called Black mecca, it is notable that the city's eventual shift by no means dismantled many white Southerners' resistance to racial desegregation and unwillingness to embrace social progress entirely. This situation is best exemplified in the nagging persistence of "the plantation" as a perennial site of popular tourism and cultural memory in the metro Atlanta area. Captured most famously in both the literary and film versions of *Gone with the Wind*, Atlanta's outsized legendary status as a renowned city largely rests on the resilience of the white supremacist culture of the slaveholding planter elite dominant during antebellum South era. Atlanta expanded dramatically both immediately before and throughout the duration of the U.S. Civil War. Despite being such a relatively "new" city within the antebellum American context, its strength becomes quickly clear: while other cities still thrive on shipping and agriculture, Atlanta rapidly modernized and adapted itself to support the Southern slavocracy's war mobilization effort by rapidly becoming the manufacturing and industrial hub of the Confederacy. However, from the names of residential subdivisions and major thoroughfares throughout the city to the large number of "plantation-themed" destination weddings and historic tours, the myth and mystique of the Old South has been a thriving cottage industry in this "New" South despite (or possibly because of?) its decades of Black political leadership and economic clout. In many respects, it is staggering to witness how the perceived gentility and grandeur of plantation culture has persisted quite doggedly alongside the so-called "religion of the lost cause" within the contemporary white Southern racial imaginary.[5] Ultimately, this mythic vision of Atlanta has represented a tacit complicity of a certain contingent of white natives and tourists alike to reclaim the city's failed promise as a sort of Confederate utopia either overtly or covertly.

Emerging from the murky haze of Jim and Jane Crow segregation, Atlanta's local power brokers during the peak of the civil rights era reassured wary capitalists from far and wide that, unlike the tumult tearing apart many cities throughout the U.S. South in their transition from grotesque racist acrimony to greater racial harmony, their city was still safe for business. Wealthy white moderates, Black leaders, and white officials agreed that profit and progress didn't have to be mutually exclusive—if the city gradually desegregated in an incremental fashion and remained economically stratified, then full-blown racial strife could be kept at bay.

Despite best intentions, this blithely cheerful albeit ahistorical image of the city hid a deep-seated pattern of racist terrorism grated onto neoliberal policies. As historians Andrew J. Diamond and Thomas J. Sugrue have argued, "In the United States, it has been impossible to separate structural arguments about markets and privatization from moral, political, and cultural frameworks that create, reinforce, and perpetuate racial ideologies and inequalities."[6] For instance, white homeowners went to extreme lengths to avoid having Black neighbors, including threatening prospective buyers and even, on several occasions, bombing their own homes. Through the 1970s and beyond, droves of whites fled this feared incursion for suburbia, convinced that a Black takeover was imminent. In fact, there were so many white residents who fled Atlanta for the suburbs during the 1960s and 1970s that the city earned another, even more poignant nickname: "the city too busy moving to hate."

Challenging the conventional wisdom that white flight meant nothing more than a literal movement of whites to the suburbs, this essay argues that it represented a more important transformation in the political ideology of those involved. In *White Flight* (2005), historian Kevin Kruse provides a critical reappraisal of racial politics in modern America in which he explains that segregationist resistance, which failed to impede the civil rights movement, prompted widespread white suburbanization in Atlanta and elsewhere. Seeking to understand segregationist thought and action on its own terms, Kruse moves past simple stereotypes to explore the meaning of white resistance to preserve the world of segregation and even perfect it in subtler and stronger forms. In a provocative revision of U.S. history of the post–Second World War era, Kruse demonstrates that traditional elements of modern conservatism, such as animosity toward the federal government and faith in free markets, underwent important transformations during the postwar struggle over segregation. Likewise, white resistance gave rise to several new conservative causes, like tax revolt, charter schools, tuition vouchers, and privatization of public services. Within the context of modern American politics, Kruse identifies how the sociopolitical journey of southern conservatives infused their vision of white supremacy into the origins of white suburbia. In accounts of the rise of modern conservatism and the New Right, a range of political languages and ideas—consumer rights, homeowners' and property rights, meritocracy, entrepreneurialism, individualism, freedom—were crowded under the umbrella of "white flight." Moreover, seen from another perspective, this phenomena also could be read as symptomatic of the cross-fertilization of white supremacist viewpoints and neoliberal values within U.S. political culture.[7]

Beginning with the election of Maynard Jackson in 1974, Atlanta has had an uninterrupted streak of Black mayors. This has been the most visible emblem of Atlanta's rebirth as a Black mecca and a global city. Nevertheless, to the astonishment of many, Black stewardship brought prosperity. Black cultural and educational institutions became ascendant, and Black wealth was more commonplace there than in almost any other American metropolis. As with other metropolises of comparable size and status during this stage of late phase capitalism, Atlanta has escaped the urban death spiral of the past half century associated with the decline of older, more traditional agricultural and manufacturing industries by largely reinventing itself around knowledge and tech-based industries like banking/wealth-management services as well as lucrative forays into music, television, film, and sports entertainment. While the United States is becoming more diverse, the opposite has been happening in many American cities over the past two or three decades. During this period, numerous city governments have invested sizable sums of money into public infrastructure projects such as parks, transportation hubs, eateries, entertainment venues, and the like in order to attract more residents even as developers have built new upscale residential buildings.[8] In *Beyond Plague Urbanism*, Andy Merrifield documents how modern metropolises globally had been plagued by social injustices and economic inequalities long before COVID-19 upended urban life everywhere. Merrifield delves into this zone of urban pathology and asks what successive lockdowns and exoduses, remote work and small-business collapse, redundant office space and unaffordable living space portend for our society in cities.[9] Even looking beyond the trauma of the Trump years and the tragedies incumbent to the COVID-19 pandemic, the process of American urban deindustrialization has been long and uneven.

Terms such as "deindustrialization" and "postindustrial" are deeply, innately contested since most cities continue to serve as home to trade and manufacturing on some level. Yet between the end of World War II and the early twenty-first century, cities that once depended solely upon manufacturing as their lifeblood increasingly began to diversify their economies when faced with the larger global, political, and demographic transformations of this historical epoch. Manufacturing centers in New England, the mid-Atlantic, and the Midwest United States were soon identified as belonging to "the American Rust Belt." Meanwhile, Atlanta proudly claimed its primacy as the proverbial capital of the "New South." Steel manufacturers, automakers, textile fabricators, and countless other industrial behemoths that once dominated urban landscapes as economic and sociocultural mainstays closed their doors as factories and workers followed economic and social incentives to leave urban cores for suburbs, the U.S. South, the Sunbelt, the Pacific Northwest, or even traveling offshore to foreign countries thanks to global trade agreements. Remaining industrial production became increasingly automated, resulting in significant declines in the number of factory jobs. Metropolitan officials faced with declining populations and tax bases responded by adapting their assets—in terms of workforce, location, or culture—to new economies, including warehousing and distribution, finance, health care, tourism, leisure industries like casinos, and privatized enterprises such as prisons. Faced with declining federal funding for renewal, they focused on leveraging private investment for redevelopment. Deindustrializing cities marketed themselves as destinations with convention centers, stadiums, and festival marketplaces, seeking to lure visitors and a "creative class" of new residents. While some postindustrial U.S. cities such as Atlanta clearly have become success stories of reinvention, many others have struggled to escape oblivion.[10]

While downtown areas in U.S. metropolises were seriously challenged prior to the COVID-19 global pandemic, their fate in the post-pandemic world has left them as "wounded renditions of their once-robust selves."[11] Years after the pandemic lockdowns closed many workplaces, entire office suites remain vacant and gathering dust. Empty commercial real estate is steadily losing value, which in turn is pulling down the municipal tax revenue for many cities nationwide. In the COVID-19 pandemic's earliest phase, the presumed duration of "remote work" was predicted to last only a few weeks or even a couple months. But currently it seems abundantly clear that a hybrid workplace with some modified combination of in-person and remote work will have an inevitable impact on urban life for years to come. Additionally, the decreased volume of daily foot traffic by fewer downtown shoppers during the pandemic accelerated an already pronounced overall downturn trend in employment within the urban core. In writing about the future of cities, author Dror Poleg describes how the evacuation of occupants from downtown office buildings will invariably have knock-on transformative effects on the larger political economy of urban America concerning finance, work, real estate, and technology.[12] Although many public mass transportation systems have gravitated toward somewhat costly yet greener technologies, many municipal governments also face deep fiscal challenges because fewer commuters using mass transit on a routine basis equates to less transit revenue to reinvest in more ecologically responsible infrastructure. The cumulative impact of these interrelated factors has led some urban planners and economists to worry about the impending challenge of an "urban doom loop" potentially unravelling the social fabric of modern America.[13]

As cities continue wrestling with myriad problems associated with life in the post–COVID-19 pandemic era, urbanists Edward Glaeser and Carlo Ratti in the *New York Times* propose a new paradigm for urbanity, namely the "playground city." On a global scale, it has been imagined since the 1970s that the pathway of dynamic economic growth and successful urban revitalization vis-à-vis street-level, seamless public infrastructure—restaurants, scenic parks, theaters, music halls, and public squares—would be through sites of pleasure.[14] When describing what the idea of the playground city is, Glaeser and Ratti say, "the Playground City differs from the industrial or the office city because it is focused on the jobs of everyday life." Furthermore, they contend future visions of cities ought to embrace a transition "from vocation to recreation" and make cities more vital and fun in which to live.[15] Cities should "attract the rich and talented," reconfiguring neighborhoods into walkable safe spaces with lots of amenities.[16]

While the model of the playground city has some useful and even utopian ideals embedded within it, Glaeser and Ratti's analysis ignores the endemic nature of sociocultural, economic, and structural problems rooted in historical inequity confronting urbanization, which the "playground" concept could make worse. For example, viewed in the context of Atlanta as playground city in tandem with the creation of Cop City, perceived crises have included the threat and reality of mass uprisings against police violence, extreme and racialized income inequality and displacement, corporate media narratives in the wake of the 2020 uprisings that threatened the image of the city as a safe place for capital investment and development, and a municipal secession movement—white flight—that threatened to rob the city of nearly half of its tax revenue following the uprisings. Atlanta as a "city of forests" would be transformed/deformed from a woodland paradise into a playground city of office parks, luxury condos, and shopping malls in accordance with the "right to comfort" for white citizens. In their 2016 workbook *Dismantling Racism*, Kenneth Jones and Tema Okun provide a lucid definition of the "right to comfort" as an internalized mode of white supremacy stated as "the belief that those with power have a right to emotional and psychological comfort. . . . I have a right to be comfortable, and if I am not, then someone else is to blame."[17] In her article, "The Cost of White Discomfort," Brittany Packnett Cunningham writes:

> Entering public spaces all but guarantees some discomfort in a country unashamed of its violently ineffective systemic responses to housing and mental health crises among its residents. There, each of us, whether we care to or not, are forced to come face to face with the unkept promises of the democratic experiment. The privileged get to merely observe the nightmare of poverty and despair instead of live it, and when their discomfort grows too great, they take matters into their own hands.[18]

From Backwoods to Backstreets: Metro Atlanta's Urban-Rural Divide and the Racial Geopolitics of Urban Space

By late May 2020, barely two months into the global COVID-19 pandemic, a discernible shift in natural phenomena had swept over the city. For instance, when stay-at-home mandates were finally imposed, the once congested highways and byways of one of the nation's most overcrowded cities were emptied of any sign of traffic for days on end. Consequently, the dense layer of smog hovering overhead had begun to dissipate, and residents could see the whitening of the clouds with the naked eye. In addition, many residential green spaces, backyards, and other wooded terrain had transformed quickly into impromptu wildlife preserves wherein deer, raccoons, wildcats, and even coyotes roamed freely with bored indifference toward humans while sizable flocks of birds canvassed the skies overhead with a bravado that would put the cinematic musings of Alfred Hitchcock to shame.

Like other cities around the nation, office space in Atlanta's dense downtown areas even several years after the pandemic's peak impact remains below full utilization, as workers spend more days working from home and many businesses struggle to adapt to this new commercial paradigm. The same goes for weekday ridership on the city's MARTA public transit system, previously one of the more renowned and reliable in the southeastern U.S. region. What happens to American cities if they are no longer the place to which people commute each weekday? Undeniably, there's a great deal that city leaders can learn from each other about reimagining cities during this latter phase of the pandemic. Considering these factors, this essay ultimately seeks to explore the great and growing need for civic leaders, urban planners, community organizers, and other major stakeholders to discuss some of the key problems Atlanta and other cities have faced during pandemic recovery; negotiating actionable solutions, the possible role of federal intervention, the uniqueness of interrogating and interpreting this post–COVID-19 pandemic and post–BLM protest moment in history become all the more important.

Considering NASA's recent determination that the scale and scope of the global heat waves during the summer of 2023 make this the Earth's hottest year on record since at least 1880, it may be useful to discuss how Cop City will lend to worsening the "heat island" effect in the metro Atlanta area. As defined by the U.S. Environmental Protection Agency, the "heat island" effect is a climate disaster wherein urban areas experience higher temperatures than outlying areas because hard, dry human-made structures such as buildings, sidewalks, roads, parking lots, and other infrastructure absorb and reemit the sun's heat more than natural landscapes such as lawns, forests, and water bodies. Moreover, urbanized spaces, where these inorganic structures are densely concentrated and green spaces are increasingly scarce, become islands of hotter temperatures relative to outlying areas. Included in the EPA's list of factors that contribute to the creation of heat islands are reduced natural landscape in urban areas, the inherent properties of urban building materials, the geometry of urban landscapes, and heat generated from human activities. There's no doubt the construction and daily operation of the Cop City facility will involve all the aforementioned factors that, in turn, only will hasten the impact of the heat island effect's higher temperatures, less shade, and reduced moisture on Atlanta within coming years.[19]

Although this is not an explicit dimension of the Stop Cop City movement, there needs to be consideration about how the unfettered spending on policing vis-à-vis Cop City is diametrically opposed to approaches to public health care in metro Atlanta. For instance, when the growing outbreak of the novel coronavirus made its way to the United States in early 2020, Atlantans were dying by the hundreds at that point and numerous sectors of the global economy ground to an absolute halt. Atlanta mayor Keisha Lance Bottoms got into a very public fight with Governor Brian Kemp over the stay-at-home order and mask mandate she wanted to impose, despite Kemp steadily receiving urgent warnings from various public health officials, most notably from Centers for Disease Control and Prevention, which has its headquarters in Atlanta.[20] Eventually even Mayor Bottoms and her family had to quarantine themselves when they tested positive for COVID-19.

The *Atlanta Journal-Constitution* editorial board highlighted how the closure of the Atlanta Medical Center is merely one example of the Georgia public health-care system failing under Gov. Brian Kemp and risking Georgians' health and well-being. The Atlanta Medical Center was the sixth hospital to shut down under Kemp's gubernatorial tenure even amid the active menace posed by COVID-19. And yet Kemp still adhered to austerity measures deeply popular among conservative ideologues that ultimately led him to refuse expanded Medicaid, thus preventing Georgia from accessing billions in federal funds to help support struggling municipal and state hospital systems and blocking five hundred thousand eligible Georgians from accessing affordable health care. Moreover, the *AJC* editorial board piece also notes that, under Kemp and his fellow Republicans, nearly 1.4 million Georgians have no health insurance, an uninsured rate of 14.5 percent, the 2nd highest in the nation and 50 percent higher than the national average.[21] However, community advocates and health policy experts say that the hospital closures in Atlanta exemplify systemic inequality that will disproportionately and most deleteriously impact low-income and BIPOC (Black, Indigenous, People of Color) communities. The closures are not unique, but rather they are part of a much larger pattern of urban hospital closures across the United States. Over the past few decades, rural hospitals also have been disappearing at an alarming rate. Notably, inner-city health-care facilities in Chicago, Philadelphia, and Washington, DC, also were shuttered circa 2019. However, at the outset of the COVID-19 pandemic, residents in the surrounding poor, predominantly BIPOC communities were left reeling from the cumulative loss of these key bulwarks of the civic healthcare safety net.[22]

Likewise, Wellstar Health System closed its Atlanta Medical Center by November 1, 2022. According to the Lown Hospitals Index for Social Responsibility, Wellstar's Atlanta hospital was ranked the second best healthcare provider in Georgia for patient inclusivity, based on the local community's demographic metrics such as racial, socioeconomic, and educational levels. Wellstar Atlanta Medical Center was one of few hospitals caring for a disproportionately high number of patients from lower-income communities and communities of color, making this facility an outlier in arguably one of the most racially segregated hospital markets in the nation. Without question, hospital closures of this sort had a considerable impact on the metro Atlanta area. Losing one of the most inclusive, socially responsive hospitals statewide in this fashion posed even greater challenges to the region's perennially fragile public health infrastructure. Most germane to this discussion, Wellstar Atlanta Medical Center faced one of the highest COVID-19 burdens when upwards of 10 percent of all its hospital beds were occupied by COVID-19 patients for nearly the entire first year of the pandemic. Comparably, other nearby hospitals passed this ominous threshold during less than half the time they did. As current Atlanta mayor Andre Dickens commented, this wave of hospital closures "will have deep and reverberating consequences for the half a million residents of Atlanta and the hundreds of thousands of visitors and commuters to our city each day."[23] Despite the selective outrage of the city's panicked and paranoid white elites, it seems evident that a better path to public safety in metro Atlanta might depend more on expanding the city's public health infrastructure instead of building Cop City.

The role of infrastructure development in geopolitical concerns of local public health and ecological factors is likely to be at the core of new forms of strategic contestation among citizens, corporations, and the state across the coming decades. This is a rare historical moment, one where the rise of a great power coincides with a wave of global infrastructure building and city shaping. Without question, the infrastructural patterns laid down today will shape the lives of millions of people over the coming decades, if not centuries. This moment connects geopolitical power, economic relations, ideological visions, and material forces in ways that presage and configure a weltanschauung as well as a historical epoch, which is why the foothold established by the Stop Cop City activists is so vitally important.[24]

Moving away from loosely defined urban theories and contexts, the racial geopolitics of urban space approach argues that it is time to start learning from and compare across different "contested cities." It questions the long-standing Eurocentric academic knowledge production that is prevalent in urban studies and planning research. Targeting previously ignored stories or misconstrued data, racial geopolitics of urban space as an analytical frame offers an in-depth understanding of the worldwide contested nature of cities in a wide range of local contexts. Even more, it suggests an urban ontology within the U.S. context that moves beyond the urban West and North as well as adding a comparative-relational understanding of the contested nature that Southern cities such as Atlanta are developing in a literal sense. To be sure, it is quite probable that Atlanta will be more exemplary than exceptional in this regard.

By way of illustration, Jonathan Rokem and Camillo Boano's edited volume *Urban Geopolitics: Rethinking Planning in Contested Cities* (2018) seeks to further define and explore the expanding world of cities across interconnected scales, moving away from a deterministic view of the incommensurability of cases from diverse regional settings and extending the comparative research of urban difference.[25] A key attribute governing the geopolitics of urban space more generally is that it is not a theory-based approach—there is no singular urban geopolitical grand theory or theories per se. Instead, the geopolitics of urban space focuses on "quality of life" issues via localized case-study analysis with a present-oriented, grounded (pun fully intended) outlook while remaining keenly aware of "glocal" (the fusion of global and local) realities of contemporary urban life with the interactions of international relations and statecraft. Inevitably, it uncovers the mundane, messy technologies of power relations in cities and their connections with global, national, and regional scales of political violence and spatial control.[26] On another level, the geopolitics of urban space as an emerging field of interdisciplinary study aims to bring geopolitics into the mainstream of urban studies to enhance an understanding of cities as contested nexus points of social, spatial, and political change. Underlying the field's various contributions is the argument that different kinds of contested cities are increasingly similar due to ethnic, racial, and class conflicts revolving around issues of housing, infrastructure, participation, and identity, among others. As both a discipline and a discourse, racial geopolitics of urban space adopts a postcolonial comparative and relational perspective while it calls into question the long-standing Eurocentric approach in urban studies that focuses on methodological regionalism and assumes incommensurability as well as asymmetrical relations.[27]

If You Love This City: Black Civic Leadership, Black Lives Matter Protests, and the Genesis of Cop City

The elements of the Stop Cop City movement—the histories upon which it draws, what it is fighting against and for, and whom it is bringing together as well as how—have given it tremendous staying power despite extraordinary odds. Local leaders as well as residents often proclaim Atlanta as "a city in a forest," with a thick canopy of trees covering nearly half of the land. The ecosystem depends on this foliage, and activists say that the deforestation required to build the facility will harm air quality, hasten climate change, and contribute to flooding in predominantly poor and working-class Black and brown communities. Once upon a time, to boost its profile as a city of the future, the metro Atlanta government's urban planners and promoters bragged about the city's four vast woodland areas as the "lungs of Atlanta."[28] In addition to rupturing arguably the largest uninterrupted green spaces in Atlanta, the proposed development of Cop City will force residents into closer proximity to toxic pollutants, waste, and even chemicals associated with the detonation of explosives and other munitions that would be routinely used on the site, which will likely degrade the air, water, and land on which myriad living creatures depend. Moreover, the Stop Cop City activists argue that unchecked police violence itself constitutes an environmental hazard detrimental to the overall welfare of civilians.

While Cop City's scale and ambition are remarkable enough, the project also evokes the many public dramas regarding the place—both literally and figuratively—policing inhabits within major metropolitan areas of the nation. It is transparently a reaction to the urban uprisings of 2020, prompted by George Floyd's murder, as well as the subsequent demands to "defund the police" by many activists and organizers of the progressive Left. Considering this, Cop City must be viewed just as much as a direct rebuke to this recent political activism around police brutality as it is an effort to stem the rising tide of crime in a quotidian sense. So, while the facility is extraordinary in its scope and potential impact, the construction of Cop City also fits quite neatly into the conventional albeit unrelenting logic that the only means to foment "law and order" is via the expansion of police funding and police power. It functions further to allay the fears of Atlanta's wealthy Buckhead district, home to much of the city's white corporate and political elite. One must keep in mind that, fearing the rest of the city's alleged deterioration during the multiple crises of 2020, Buckhead's well-heeled denizens openly threatened to secede to form their own municipality. Thus, the establishment of Cop City promises to assure the city's affluent residents and real estate developers that their property will be protected should anything remotely like the 2020 Black Lives Matter protests happen ever again.

In many ways, a critical flashpoint for this conflict occurred on a sunny Wednesday in November 2020 when Atlanta mayor Keisha Lance Bottoms walked into the Atlanta City Detention Center. More than a generic press event, Bottoms visited this facility with the intention that it would serve not merely as a photo op but also as a victory lap showcasing her administration's accomplishments prior to her departure from city hall. Standing at a podium inside the hulking, water-stained concrete structure that looms over Peachtree Street with slits for windows and floors consisting of a threadbare mix of shabby linoleum and run-down brown carpet, the mayor stared tearfully at those gathered in the thoroughly well-worn building. Mayor Bottoms would later reveal that her administration successfully forged a deal remaking a portion of the building into a "Diversion Center," which, according to the facility's website, provides "immediate services instead of arresting people for activities related to extreme poverty, addiction, or mental health 24 hours a day, 7 days a week" and thus "help reduce Atlanta and Fulton County's reliance on jail for public order issues stemming from quality of life concerns." Most impressively, the advocates for this new center predict that this operation "will reduce the number of people struggling with these issues from entering the city and county's jails" with "the potential to divert 10,500 jail bookings (10% of bookings) annually from the Atlanta City Detention Center and Fulton County jail."[29] From the city's first mayor, Moses Formwalt, to the pioneering generation of Black mayoral leadership by the likes of Maynard Jackson and Andrew Young, Atlanta has had many notable mayors throughout its history who have led Atlantans through growth and change over the years.[30] To what extent does Atlanta genuinely serve as an exemplar of "New South" urbanization (race relations, regional prominence, real estate, resource management, racial realignment of electoral politics, etc.)? What does more than a half-century of Black mayoral and police leadership in Atlanta mean within a white conservative Southern regime?

In the wake of Mayor Bottoms's inauguration to the highest elected office in America's so-called Black mecca in 2018, Governor Brian Kemp and his fellow GOP leadership in Georgia's state legislature barely gave her time to breathe before launching multiple attacks on her administration. One of the earliest threats was an effort to pass legislation for the state of Georgia to take control of the Hartsfield-Jackson Atlanta International Airport, the city's crown jewel. That effort disappeared once it became clear that Republicans lacked the political will for the protracted battle such an endeavor would entail. But much of her time in office was also marred by a series of unprecedented and unparalleled cries: She faced a cyberattack against the city,[31] a continued federal investigation into her mayoral predecessor, Kassim Reed, a rapidly growing global pandemic, and an increasingly volatile, potentially violent climate surrounding the 2020 presidential elections.

The summer of racial justice protests was further intensified when twenty-seven-year-old Rayshard Brooks was shot and killed in the parking lot of a Wendy's restaurant by an Atlanta police officer on the evening of June 12, 2020. More than anything else, Bottoms had been castigated publicly for how she responded to the city's unexpected rise in violent crime with the onset of the global COVID-19 pandemic.[32] It turned out that the killing of George Floyd in Minneapolis two weeks earlier would escalate the challenges facing one of the most competent, nimble, and lucky big city mayors in recent memory. Though Bottoms was able to circumnavigate a wide array of potentially massive pitfalls while in office, it was the issue of criminal justice that most grievously frustrated her political fate. Promoting the notion of Atlanta as a haven for Black excellence in the heart of the Deep South was also the leitmotif of her political ascent. In the summer of 2020, however, this once exceptional city looked just like the rest of the nation: an emotionally beleaguered and morally outraged populace at war with itself. After the online video of George Floyd's murder by four Minneapolis police officers went viral around the globe, there were contrasting scenes on display during the twenty-four-hour news cycle of peaceful protesters marching by the tens of thousands proclaiming "Black Lives Matter" on the one hand, and raucous rioters smashing downtown storefront windows and setting police cars ablaze on the other. When grappling with this in her beloved hometown, Mayor Bottoms was visibly shaken by this turmoil as she took to the media airwaves to deliver a speech intended to calm her fellow Atlantans. "This is not a protest. This is not in the spirit of Martin Luther King Jr. This is chaos," Bottoms said. Furthermore, the mayor stated, "When Dr. King was assassinated, we didn't do this to our city. So, if you love this city—this city that has had a legacy of Black mayors and Black police chiefs and people who care about this city, where more than 50 percent of the business owners in Metro Atlanta are minority business owners—if you care about this city, then go home."[33] During her tenure in office, Bottoms's time on the national stage was defined by both soaring political heights that placed her prominently on the shortlist as Joe Biden's potential Democratic running mate in the 2020 election and staggering lows, most especially her surprising May 2020 announcement that—after a sole term as mayor of one of America's most popular and populous cities—she would not seek reelection.

That mythic nature of the metropolis hailed as the Black mecca had been defined by these two diametrically opposed worlds—the downtrodden Atlanta and the dynamic Atlanta. The debates swirling around the genesis of Cop City highlight the utility of framing environmental justice, racial hypersegregation, and criminal justice reform as issues that are racially coded as "Black," especially given the historic role the city's Black mayors play in maintaining the status quo in favor of neoliberal economic policies and consolidating state power. To better fathom the extent of this current predicament, we need to grasp what philosopher Achille Mbembe calls necropolitics, which he describes as "the ultimate expression of sovereignty [that] resides . . . in the power to dictate who may live and who must die."[34]

The Black Lives Matter movement has declared an outright crisis of domestic life within the United States and has demonstrated it has extraordinary potential to transform American society.[35] While the movement has grown into a more powerful and organized form that faces extraordinary challenges since it began over a decade ago, BLM insists that racialized acts of violence are neither coincidental nor accidental in nature. Far from merely being the misconduct of a few police officers "acting badly," advocates and allies of BLM denounce the purposeful use of force—including lethality—at the behest of the state for the protection of the wealthy and preservation of property that has been incumbent to policing in America since its origins. The formation of public and private police from the colonial era to the outset of the U.S. Civil War has roots in racialized social control and colonial violence related to the control, containment, and capture of enslaved people as well as the protection of property and infrastructure for emerging planter elites.[36] Ruth Wilson Gilmore's widely cited definition of racism as "the state-sanctioned or extralegal production and exploitation of group-differentiated vulnerability to premature death" reminds us that the United States has always been a deadly force in the lives of Black people.[37] Scholarly discourse regarding the militarization of police, state and corporate surveillance, the carceral state, perpetual war, and the "state of exception" is hardly new for understanding the existential concerns of Black people, yet the recent developments since 2020 have been fraught with greater exigencies about positionality, privilege, and power.[38]

Who Watches the Watchmen? Policing the Black Mecca in the Post–COVID-19 / Post–BLM Era

Crime was the top concern for most metro Atlanta voters ahead of the November 30, 2020, runoff mayoral election to find a successor for Mayor Bottoms. Since the wave of BLM protests in 2020, Atlanta voters had become transfixed by crime. Homicide rates in the city increased by 50 percent from 2019 to 2021. Accordingly, the leading candidates—Atlanta City Council president Felicia Moore and Councilmember Andre Dickens—were far to the right of Bottoms on the issue of policing. While those statistics were roughly on pace with Chicago and Philadelphia, and the data could be read as part of a national trend, they were still a statistical aberration amid a historic lull that naysayers could use to their maximal benefit. As the more pacifistic of the two runoff candidates, Dickens triumphed and became the sixty-first mayor of Atlanta.[39] Dickens was sworn into office in January 2022 and shared his vision for Atlanta in his first State of the City address wherein he envisioned:

> One city with one bright future. A city of safe, healthy, connected neighborhoods with an expansive culture of equity, empowering upward mobility and full participation for all residents, embracing youth development, and an innovative, dependable government moving Atlanta forward, together.[40]

His administration claimed that early accomplishments of his "Moving Atlanta Forward" agenda included making the city's first-ever investment in early childhood education, the establishment of the Nightlife Division to address establishments with a history of high crime, the reestablishment of the Pothole Posse to rapidly respond to residents' reports, and leading the successful coalition to keep Atlanta whole in opposition to a de-annexation effort by some of its more monied residents.[41] Not surprisingly, Dickens attempted to solidify his political fortunes by increasing opportunities for the city's young people, empowering neighborhoods, and investing in housing and combating homelessness, all while fostering a culture of integrity in the city with promises of improving public safety.

Without a doubt, the geopolitics of urban space shifted incredibly due to the global COVID-19 pandemic. Most of the plans for the Cop City facility were sparked by a startling uptick in major crimes coinciding with the onset of the COVID-19 pandemic not only in the metro Atlanta area but in countless other cities across the United States. Considering these factors, there is a great deal that city leaders, urban planners, community organizers, and other major stakeholders can learn from each other about reimagining cities like Atlanta during this latter phase of the pandemic. But even as the worst of both the pandemic and crime surge subsided, some of the more positive changes have yet to take root. For example, crafted in 2020 following a massive groundswell of BLM protests globally, the George Floyd Justice in Policing Act was intended to address structural racism, racial profiling, pervasively aggressive paramilitary police culture, and the excessive use of deadly force in law enforcement among other things. But the bill still has not been passed into law by the U.S. Congress, and it is unclear if it ever will. This political impasse invokes the question memorably posed by the Roman writer Juvenal millennia ago: *"Quis cust-diet ipsos custodes?"* (Who watches the watchmen?). Thus far, progress has often been stymied by conflicting ideas across the political spectrum about the prospects for policing reforms in an arguably post-COVID-19 world as well as how such efforts could be instructive for broader social transformation, particularly the binary "us vs. them" mentality prevalent in so many policymaking and partisan political circles.

For several decades now, Republican politicians and broader rightwing rhetoric have obsessed over the state of crime in the United States and use any perceived rise in violent incidents to mobilize their electorate based on fears of lawlessness. In like fashion, violent crime served as a huge focus for GOP candidates during the 2022 midterm elections. Republicans spent about $50 million on crime ads for the two months leading up to those elections. This ad campaign largely pushed a dystopian vision of cities flooded with graphic scenes of murder, robbery, and assault, with Republican politicians predictably attacking their Democratic rivals for presumably being unwilling or unable to act.[42] The Brennan Center for Justice reported that the number of murders per one hundred thousand people rose by nearly 30 percent nationwide in 2020, while aggravated assault also rose by 11.4 percent. The murder rate rose noticeably in big cities, which typically tend to vote for Democratic candidates while also being repeatedly and relentlessly demonized by Republicans.[43] But since reaching that statistical apex, most types of violent crime have now fallen. According to the Council on Criminal Justice, crime dropped in thirty-five large cities in 2022, even though the overall rates remain higher than prepandemic levels. Nevertheless, the rate of homicide in major cities was about half that of historic peaks in the 1980s and early 1990s.[44] However, as illustrated in the Cop City controversy, this sort of political fearmongering vis-à-vis "tough-on-crime" political ads and messaging in campaign speeches that continues to saturate our media ecosystem and dominate public discourse will further exacerbate the nation's conjoined policing and mass incarceration problems in advance of upcoming elections.

The data compiled by the U.S. Crisis Monitor reveal that the United States is indeed a nation in crisis. It faces a multitude of concurrent, overlapping risks—from police abuse and racial injustice to pandemic-related unrest and beyond—all exacerbated by increasing polarization. This report mapped these trends with a keen view toward the 2020 election, when these intersecting risks are likely to intensify.[45] Since the Black Lives Matter movement surfaced as a visible and vibrant mass mobilization of protesters in 2014, it has been quite credibly compared to the U.S. civil rights movement of the 1950s and 1960s.[46] Whereas BLM advocates and allies have argued that this movement is clearly a continuation of previous efforts to advance social justice within a contemporary context, the movement's critics have argued expectedly that the BLM movement has lacked the U.S. civil rights movement's dogged emphasis on nonviolence *qua* Kingian pacifism as its central organizing principle. After the alleged "summer of racial reckoning," conservative media outlets have condemned BLM-inspired protests for "tearing apart our cities" and argued that regardless of their intent or impact, the 2020 protests were a poor imitation of those held during the civil rights era.

Contrary to such negative claims, recent research has found that the 2020 BLM-related protests were in fact overwhelmingly peaceful. Political scientists Erica Chenoweth and Jeremy Pressman reported that their Crowd Counting

Consortium (CCC) found that less than 4 percent of the BLM protests involved property damage while 1 percent involved police injuries. Other data collections similarly found that 95 percent of the protest marches were peaceful. Moreover, when compared to the civil rights era protests circa the 1960s, their research findings revealed that on every available metric, the 2020 BLM protests were more peaceful and less confrontational. As a case in point, civil rights protests and demonstrations were most widespread from 1960 to 1968. From the student sit-in in Greensboro, North Carolina, on February 1, 1960, to the assassination of Rev. Dr. Martin Luther King Jr. on April 4, 1968, civil rights activists took coordinated acts of civil disobedience via nonviolent direct action against racial discrimination and other forms of social inequality. King and other movement leaders argued that protesters' nonviolence would contrast with their opponents' violence while also providing tremendous moral authority, even while critics blamed them for instigating the violence that was being inflicted on them. The CCC's findings indicate that, while the civil rights movement protests were primarily nonviolent (at least for the protesters), the BLM protests were proven to be even more so. According to Chenoweth and Pressman's research, 11 percent of the 2,681 events in the 1960s contained property damage during these eight years of civil rights protests. By contrast, during the 12,839 racial justice protests in 2020, only 4 percent involved property damage. Also, police officers were injured in 6 percent of the civil rights movement's protests, but police officers were reported injured in only 2 percent of the 2020 protests. Conversely, police officers were also more openly aggressive during the civil rights movement. Police arrested some protesters in 36 percent of all civil rights movement events, but the arrests in the recent BLM protests were just 7 percent. Moreover, as many know, while attacking civil rights marchers on the 1965 Selma-to-Montgomery march in Alabama that infamously became known as "Bloody Sunday," police officers bludgeoned John Lewis and fractured his skull. Consistent with that, civil rights era protests were more dangerous for protesters than the 2020 protests; for instance, the CCC report found that there were crowd injuries from 12 percent of all civil rights movement protests, compared with only 3 percent of the 2020 protests.[47] By and large, the accusations of violence and lawlessness associated with the 2020 BLM protests were grossly overexaggerated.

The largest group of arrests, on March 5, 2023, was in a public park in the forest near where the Cop City project is planned. This police action was followed promptly by local government closing the park, in effect shutting down the protests of tree-sitting "forest defenders" that had gone on for more than a year as of this writing.[48] While the Stop Cop City movement is a welcome intervention, it nevertheless does not suggest that there is innovation in the operation of anti-Black violence in our present postpandemic era, but this innovation belies the fact of exclusionary public policies or a militarized police force consolidating within urban life and culture. There must also be an awareness that there are dynamic demographic shifts in the social order that policing actively supports, which are assembled through the racial geopolitics of urban space. A burgeoning scholarship on race, political economy, and urban space lends itself to the broader contemporary thought process wherein we find insights into historical debates about "internal colonialism."[49] To be sure, this literature explicitly locates the "Black ghetto" as a space of "colonial administration" within the domestic territory of the U.S. empire, emphasizing at once the geopolitics and political economies of racial capitalism. These debates happened largely beyond the bounds of academic scholarship yet offer a valuable racial analysis of geopolitical urban space. Similarly, it is possible to bear witness to transformations in the domestic face of American empire—shifting geographies of Black dispossession through gentrification and the subprime mortgage crisis that triggered the Great Recession of 2008–10—to bring more attention to the relationship between the economic recklessness evident in the U.S. domestic housing market and U.S. imperialism's heavy investment in the global war on terror overseas. This realization is deeply aligned with Keeanga-Yamahtta Taylor's yearning to see a "broader interrogation of American society" than a static focus on policing allows, yet we see anti-Black police violence as a fulcrum of empire from where such interrogation begins.[50] It is precisely this kind of interrogation that seems to be increasingly explicit within the inchoate yet insightful spatialized politics, antisystemic perspective, and coalitional praxis of the Stop Cop City movement.[51]

As the Cop City controversy intensified, two Democratic U.S. senators representing Georgia, Jon Ossoff and Raphael Warnock, had said little publicly about the situation until Axios reporters approached them about the matter in the U.S. Senate building's basement circa March 2023. This is a noteworthy situation because it was their proven credibility in progressive Democratic and activist circles that helped catapult both to the U.S. Congress. Prior to entering electoral politics, Rev. Dr. Raphael Warnock was best known for having earned a PhD in systematic theology from New York City's Union Theological Seminary under the tutelage of pioneering Black liberation theologian James H. Cone and served as the senior pastor of the historic Ebenezer Baptist Church previously led by Martin Luther King Sr. and Jr. as well as serving as pastor to the late civil rights icon Congressman John Lewis. Meanwhile, Jon Ossoff's election to the U.S. Senate in 2021 marks an equally important and interesting bookend event signaling the implosion of the GOP during the Trump years. As a young documentary filmmaker he used to work as a congressional staffer for Representative Lewis, who had endorsed his protégé in both elections before he passed away in 2020. In both instances, Warnock and Ossoff stood as progressive Democrats openly seeking to help American society fulfill its best governing practices as a multicultural and multiracial grassroots democracy.[52] When pressured for a response, however, both senators gave elusive answers that stopped short of siding with either the Cop City proponents or the activists. "Peaceful protest and the expression of opposition or support to a land-use plan or the siting of a specific facility somewhere is a sacred constitutional right," Ossoff said. "What is not constitutionally protected speech is Molotov cocktails." Warnock said that "protesting in the best of the American democratic tradition needs to be non-violent," emphasizing his view that "local officials there in Atlanta have to work with the citizens of Atlanta" to find a peaceful way forward.[53] Many Stop Cop City activists have criticized Senator Warnock and Senator Ossoff for not taking a stance against the project.[54] Conversely, Rep. Nikema Williams, the House Democrat whose congressional district includes the contested site for Cop City, according to the *Atlanta Journal-Constitution*, maintained she will "always stand on the side of protesters" and that she does not "buy the whole notion that you have to choose one or the other, between standing up for our law enforcement and standing with protestors."[55]

Meanwhile, Governor Brian Kemp justified the severe punishment on a message posted via the social media app X (formerly known as Twitter), saying, "Violence and unlawful destruction of property are not acts of protest. They are crimes that will not be tolerated in Georgia and will be prosecuted fully." Shortly after issuing this statement, Kemp signed an executive order declaring an official state of emergency and authorizing the deployment of a thousand National Guard troops to arrest Cop City protesters, which lasted two weeks, until February 9, 2023. It is important to note, however, that Kemp's definition of "prosecuting fully" has changed significantly since 2020. During the BLM protests that filled the streets of downtown Atlanta, most protesters arrested were taken to the city jail for offenses that included disorderly conduct, burglary, and criminal property damage.[56] Congressman Earl L. "Buddy" Carter, House Republican for Georgia's First Congressional District, issued a resolution condemning the violence by members of the Stop Cop City movement and expressing that law enforcement must be supported by the House of Representatives.[57]

In March 2023, during a Senate Homeland Security Committee hearing, Senator Ossoff worked to root out recent increases in violent crime in Atlanta and across Georgia by pressing FBI director Christopher Wray for an update on the FBI's efforts to determine key factors driving a national increase in violence. By the time of the hearing, there had been 113 homicides that year in Atlanta——16 percent higher than in the same period in 2020, and 64 percent higher than in 2019.[58] Senator Ossoff asked Director Wray to determine the root causes of increased violence and to "refine" his assessments of measures intended to keep Georgians safe, and he also asked for the results of that evaluation. Director Wray ostensibly attributed the rise in violence to the widespread impact of COVID-19, including unemployment, decreased prison sentences and corresponding inmate releases, the prevalence of illegal firearms and interstate firearm trafficking, and staffing shortages in countless urban police departments. By June 2023, however, Warnock, Ossoff, and Rep. Jamaal Bowman, a progressive House Democrat from New York, spoke out about the arrest of three board members of the Atlanta Solidarity Fund. "These tactics, coupled with the limited public information provided so far," Warnock contended, "can have a chilling effect on nonviolent, constitutionally-protected free speech activities those of us in the fight for justice have been engaged in for years." Furthermore, Warnock argued that the arrests illustrate the fears and concerns that organizers and community members have voiced on the topic of being overpoliced and underserved vis-à-vis the militarization of police in Georgia.[59] Additionally, after months of silence, Senator Warnock issued a letter to the Department of Homeland Security (DHS) expressing concern over the potentiality of activists' First Amendment rights being violated by the exceedingly questionable terrorism charges being leveled against them. In the letter, he asked DHS to clarify whether it has designated Stop Cop City activists as "domestic violent extremists," as the governor and Atlanta police department have claimed the federal department has, and to pass down that clarification to Georgia law enforcement. Even while raising these concerns, however, Warnock bent over backwards to praise the law enforcement involved in the case.[60]

Mayor Andre Dickens joined U.S. Senator Jon Ossoff to announce $30 million in infrastructure upgrades to make downtown Atlanta streets safer. Warnock, Ossoff, and Williams were all praised by the mayor for their assistance in helping to deliver $30 million through the bipartisan infrastructure law's Safe Streets for All program to improve safety for pedestrians and bikers in downtown Atlanta. "Our vision of transportation in Atlanta is safe, multi-modal and sustainable," said Mayor Dickens. "These federal dollars, combined with the commitment made by voters through our Moving Atlanta Forward infrastructure program, will transform Central Avenue and Pryor Street into safe corridors that connect the Southside and the BeltLine to our Downtown. Thank you to the Biden Administration and Sen. Ossoff, Sen. Warnock and Rep. Williams for continuing to invest in making Atlanta a city built for the future."[61]

While currently under construction, most of the plans for the Cop City facility were sparked by a startling uptick in major crimes coinciding with the onset of the COVID-19 pandemic not only in the metro Atlanta area but in countless other cities across the United States. But even as the worst of both the pandemic and the crime surge have subsided, it is probably best to envision the Cop City project as a metropolitan police academy on steroids. The complex, which will be nestled where a forest once breathed, is much more than a "police training center" but is primed to become, as Davarian Baldwin and Joshua Clover argue, a veritable university dedicated to American urban policing.[62] Easily among the largest of its kind, the Cop City campus will feature an array of learning facilities: a shooting range, a driving obstacle course, a mock cityscape for developing advanced tactics and techniques for "urban-police training," but also classrooms and an auditorium. To further illustrate, the most recent Census of Law Enforcement Training Academies (CLETA) survey conducted by the Bureau of Justice Statistics in 2018 revealed that, among the 615 academies included in the census, 39 percent of them were affiliated with a postsecondary institution and constituted the single largest category of academy affiliation. Specifically, 6.8 percent of the academies were associated with four-year colleges and universities, while 32.2 percent were associated with two-year colleges. Additionally, CLETA survey results also revealed that 50 percent of the police academies indicated an associate degree was offered through the academy, and 9.2 percent responded that a bachelor's degree was offered through the academy. To make matters worse, not only are colleges and universities at least partially funding and legitimating the degree credentials for these police academies nationwide, but there is also the fact that nearly 50 percent of the basic-training curriculum used by them is dedicated to such areas as criminal procedures, patrol tactics, strategic driving, lethal and nonlethal weapons training, martial arts, and defensive tactics as opposed to just 5 percent of curricular offerings being devoted to creative problem solving, cultural diversity, interpersonal communication, conflict de-escalation, and legal ethics.[63] Baldwin and Clover's provocative essay exemplifies why and how Atlanta's Cop City development should concern scholars to the extent police academies are always already associated with academic institutions of higher education, although we do not culturally acknowledge or fully describe them in this manner. Both implicitly and explicitly, contemporary U.S. higher education's myopia toward effectively helping facilitate the crisis surrounding police-community relations regarding matters of transparency and accountability facilitates what Caitlyn Lynch calls "the right to remain violent."[64]

The proposed police university, nestled within a major metropolitan area already home to several top-tier academic institutions, aims to provide instruction for officers in the "most up-to-date methods of community policing, including de-escalation tactics and cultural awareness."[65] While most trainees will hail from Atlanta or Georgia, it is estimated that roughly 43 percent of attending officers will come from other cities and states. According to Dave Wilkinson, president of the Atlanta Police Foundation, the nonprofit group building the facility, Cop City is being touted as "a beacon for what we call 21st-century policing." For those who somehow think Cop City will not affect them, it must be considered that at least 40 percent of police officers trained there will come from all around the nation.

In June 2023, concerned Atlantans shattered the record for participant turnout as thousands lined up en masse at a city council meeting to voice their opposition to the construction of Cop City. Ultimately, the Atlanta City Council voted 11–4 to approve, but the contentious vote occurred after a SWAT team raided the Atlanta Solidarity Fund and arrested three people who had been raising money to bail out protesters opposed to Cop City, charging them with money laundering and charity fraud. If Atlanta's prosecutors can quell the First Amendment activities of the protesters, other states might follow the state of Georgia's lead.[66] As one of the three people arrested in the SWAT raid, Atlanta Solidarity Fund organizer Marlon Kautz denounced the charges as "malicious political prosecutions" with the intent to "suppress a political movement."[67]

As with many higher-education institutions, funding for the Cop City project comes from private and public sources. When the project was originally conceived, the city was expected to pay $33 million, with the foundation and its donor network contributing the remainder of the budget. However, the initial proposed plan also indicated the Atlanta municipal government would pay another $1.2 million annually for the next thirty years earmarked to cover debt for the Atlanta Police Foundation. Therefore, it now appears that Atlanta's city hall is expected to pay as much as $67 million. It is important to note that the mounting public cost is roughly double the figure initially reported in late 2021 when the proposal was first authorized and follows more than two years of forceful opposition far and wide despite little transparency and ostensibly no recourse for Atlanta's voters. While city officials quite defensively argued that this was always the expected budget and the center's critics and opponents simply had been mistaken, it was not until a consortium of local journalists known as the Atlanta Community Press Collective audited the numbers that there had been any real effort made to rebut these accusations.[68]

Conclusion: Why the Stop Cop City Movement Matters

As indicated above, the Stop Cop City campaign builds on a history of community organizers and social justice activists challenging the neoliberal onslaught of environmental destruction as well as an effort for criminal justice reform to reframe the "defund the police" debate in quite literal and materialistic terms. This movement offers a wide range of innovative solutions to restore blighted and redlined Black neighborhoods, helping to heal communities through efforts ranging from reforestation to municipal reparations. Since urban apartheid was a sociological trend intentionally erected by human minds and hands alike, this system can be intentionally dismantled in the very same manner. Moreover, the Cop City controversy demonstrates that America cannot even begin to reckon with the reality that Black lives matter until it also recognizes how the ecology of urban Black communities and neighborhoods matter as well.

In "This Is the Atlanta Way: A Primer on Cop City," Micah Herskind writes, "Cop City is the Atlanta ruling class's chosen solution to a set of interrelated crises produced by decades of organized abandonment in the city."[69] Writing about these perceived "crises" as part and parcel of urban lived reality, Ruth Wilson

Gilmore defines the concept as "instability that can be fixed only through radical measures, which include developing new relationships and new or renovated institutions out of what already exists."[70] "It's all hands on deck for the forces of the prison industrial complex, the forces of capitalism," argues Herskind, who further states, "they are willing to use any and all tactics and tools available to them, whether that's literal murder, whether that's trying to deter the broader movement by slapping people with domestic terrorism charges." Therefore, regarding the effort to resist the construction of Cop City, Herskind notes that "as environmental catastrophe is upon us, I think the forces of capital are organizing themselves." Ultimately, the Stop Cop City movement in Atlanta illustrates an initial example of a broader spectrum of urban geopolitical provocations possibly taking shape in the not-too-distant future. This examination of the influences, issues, and impact surrounding this unlikely protest campaign outlines a new political vocabulary for modern policing and urban community relations as well as a bolder reimagining of what it means to be a product of one's environment in a more potent and profound manner. Most of all, this project concludes with an all-purpose call to broaden the theoretical frameworks, methodological approaches, and empirical datasets related to racial geopolitics of urban space to achieve a broader decolonization of current social scientific research by pursuing an agenda emanating from an interdisciplinary albeit heterodox engagement with history, political geography, urban studies, planning, architecture, sociology, anthropology, and environmental studies among others.

On the most granular level imaginable, the struggle facing the Stop Cop City movement not only has been a conflict over a large swath of woodlands in the city of Atlanta but also a battle for the future of community control of cities. It is a struggle over who the city is for: the city's corporate actors and state ruling-class elites who have demanded that Cop City be built, or the people of Atlanta who have consistently voiced their ire and opposition to being undervalued and overpoliced and thus are demanding a better, more equitable path for the city's future. It is a fight over who the city belongs to, over who is welcome to live and enjoy life here and who is expected to simply languish and labor for the diminished expectations of life, liberty, and the prospect of happiness (not merely its *pursuit*) while under constant surveillance.[71] The Stop Cop City movement as a merger of ecojustice, police abolitionism, fair housing, and economic justice movements is all those things.

NOTES

1. Rosana Hughes and Caroline Silva, "Autopsy: Gunshot Residue 'Not Seen' on Activist Killed at Police Training Center," *Atlanta Journal-Constitution*, last updated April 20, 2023, https://www.ajc.com/news/crime/activist-killed-at-police-training-center-site-had-more-than-50-gunshot-wounds-autopsy-finds/RSL5T2D7WJDXNIULLBL6ZUZADE/

2. Timothy Pratt, "'Cop City' Activist's Official Autopsy Reveals More than 50 Bullet Wounds," *The Guardian*, April 20, 2023, https://www.theguardian.com/us-news/2023/apr/20/manuel-paez-teran-autopsy-cop-city.

3. Davarian Baldwin and Joshua Clover, "Police Academies Are Part of Higher Ed," *Chronicle of Higher Education*, August 2, 2023, https://www.chronicle.com/article/police-academies-are-part-of-higher-ed.

4. Mia Gant, "Injustice Hidden Deep in Atlanta's Forest: The Old Atlanta Prison Farm and the South River," The Histories of Our Streets, April 12, 2022, https://sites.gsu.edu/historyofourstreets/2022/04/12/old-atlanta-prison-farm/.

5. Charles Reagan Wilson, *Baptized in Blood: The Religion of the Lost Cause, 1865–1920* (Athens: University of Georgia Press, 1980); Edward R. Crowther, ed., *The Enduring Lost Cause: Afterlives of a Redeemer Nation* (Knoxville: University of Tennessee Press, 2020).

6. Thomas J. Sugrue and Andrew J. Diamond, "Introduction," in *Neoliberal Cities: The Remaking of Postwar Urban America*, ed. Thomas J. Sugrue and Andrew J. Diamond (New York: New York University Press, 2020), 7.

7. Mike Davis, *City of Quartz: Excavating the Future in Los Angeles*, photographs by Robert Morrow, new ed. (1990; London: Verso, 2006); Kenneth D. Durr, *Behind the Backlash: White Working-Class Politics in Baltimore, 1940–1980* (Chapel Hill: University of North Carolina Press, 2003); Thomas J. Sugrue, *The Origins of the Urban Crisis: Race and Inequality in Postwar Detroit*, first Princeton Classics edition (Princeton: Princeton University Press, 2014); Kevin M. Kruse, *White Flight: Atlanta and the Making of Modern Conservatism* (Princeton, NJ: Princeton University Press, 2005); Richard Rothstein, *The Color of Law: A Forgotten History of How Our Government Segregated America* (New York: Liveright, 2017); Lily Geismer, *Don't Blame Us: Suburban Liberals and the Transformation of the Democratic Party* (Princeton, NJ: Princeton University Press, 2015); Keeanga-Yamahtta Taylor, *Race for Profit: How Banks and the Real Estate Industry Undermined Black Homeownership* (Chapel Hill: University of North Carolina Press, 2019).

8. Sam Bass Warner, *The Private City; Philadelphia in Three Periods of Its Growth* (Philadelphia: University of Pennsylvania Press, 1968); Arnold R. Hirsch, *Making the*

Second Ghetto: Race and Housing in Chicago, 1940–1960 (Chicago: University of Chicago Press, 1998); Jon C. Teaford, *The Rough Road to Renaissance: Urban Revitalization in America, 1940–1985* (Baltimore: Johns Hopkins University Press, 1990); Davis, *City of Quartz*; Sugrue, *Origins of the Urban Crisis*; Douglas W. Rae, *City: Urbanism and Its End* (New Haven: Yale University Press, 2003); Faranak Miraftab, David Wilson, and Ken E. Salo, eds., *Cities and Inequalities in a Global and Neoliberal World* (New York: Routledge, 2015).

9. Andy Merrifield, *Beyond Plague Urbanism* (New York: Monthly Review Press, 2023).

10. Chloe E. Taft, "Deindustrialization and the Postindustrial City, 1950–Present," *Oxford Research Encyclopedia of American History*, June 25, 2018, https://doi.org/10.1093/acrefore/9780199329175.013.574.

11. Derek Thompson, "How American Cities Can Avoid the 'Urban Doom Loop,'" The Ringer, April 18, 2023, https://www.theringer.com/2023/4/18/23687846/how-american-cities-can-avoid-urban-doom-loop.

12. Dror Poleg, "The Next Crisis Will Start with Empty Office Buildings," *The Atlantic*, June 26, 2023, https://www.theatlantic.com/ideas/archive/2023/06/commercial-real-estate-crisis-empty-offices/674310/.

13. Bruce Schaller, "The Housing Crisis' Economic Toll: The Damaging Ripple Impact from a Lack of Affordable Homes," *New York Daily News*, June 20, 2023, https://www.nydailynews.com/2023/06/20/the-housing-crisis-economic-toll-the-damaging-ripple-impact-from-a-lack-of-affordable-homes/; Tracy Hadden Loh and Hanna Love, "Breaking the 'Urban Doom Loop': The Future of Downtowns Is Shared Prosperity," Brookings, July 26, 2023, https://www.brookings.edu/articles/breaking-the-urban-doom-loop-the-future-of-downtowns-is-shared-prosperity/; Patrick Sisson, "Untangling the Urban Doom Loop," Bloomberg.com, June 21, 2023, https://www.bloomberg.com/news/articles/2023-06-21/to-break-the-urban-doom-loop-build-housing-and-transit.

14. Edward L. Glaeser, Jed Kolko, and Albert Saiz, "Consumer City," *Journal of Economic Geography* 1, no. 1 (2001): 27–50.

15. Edward L. Glaeser and Carlo Ratti, "26 Empire State Buildings Could Fit into New York's Empty Office Space: That's a Sign," *New York Times*, May 10, 2023, https://www.nytimes.com/interactive/2023/05/10/opinion/nyc-office-vacancy-playground-city.html.

16. Richard McGahey, "Cities Can't Just Be 'Playgrounds' for the Affluent," *Forbes*, May 11, 2023, https://www.forbes.com/sites/richardmcgahey/2023/05/10/cities-cant-just-be-playgrounds-for-the-affluent/?sh=1ff0394f5269.

17. Kenneth Jones and Tema Okun, "Dismantling Racism: 2016 Workbook," dRworks, https://resourcegeneration.org/wp-content/uploads/2018/01/2016-dRworks-workbook.pdf.

18. Brittany Packnett Cunningham, "The Cost of White Discomfort," *The Cut*, May 12, 2023, https://www.thecut.com/2023/05/jordan-neely-paid-the-price-for-white-discomfort.html.

19. "Learn about Heat Islands," U.S. Environmental Protection Agency, last updated August 28, 2023, https://www.epa.gov/heatislands/learn-about-heat-islands.

20. Vanessa Romo, "Georgia Gov. Brian Kemp Sues Atlanta Mayor Keisha Lance Bottoms over Face Mask Order," NPR, July 16, 2020, https://www.npr.org/sections/coronavirus-live-updates/2020/07/16/892109883/georgia-gov-brian-kemp-sues-atlanta-mayor-keisha-lance-bottoms-over-face-mask-or.

21. "Opinion: Hospital's Closing Affects Us All," *Atlanta Journal-Constitution*, September 12, 2022, https://www.ajc.com/opinion/opinion-hospitals-closing-affects-us-all/L46RPDTRZBEOLPGAL3EQBKFX3M/#].

22. "Analysis: The Loss of Atlanta Medical Center Is Part of a Larger Pattern of Urban Hospital Closures Devastating Vulnerable Communities across the US," CNN, November 12, 2022, https://www.cnn.com/2022/11/12/us/hospital-closures-race-deconstructed-newsletter-reaj/index.html; Brett Pulley and Margaret Newkirk, "Atlanta Hospital Closes in Midst of Poverty and Politics," Bloomberg.com, September 16, 2022, https://www.bloomberg.com/news/articles/2022-09-16/atlanta-hospital-closes-in-the-midst-of-poverty-and-politics#xj4y7vzkg.

23. Judith Garber, "Why Is One of Georgia's Most Inclusive Hospitals Closing?," Lown Institute, September 2, 2022, https://lowninstitute.org/why-is-one-of-georgias-most-inclusive-hospitals-closing/amp/.

24. "Cities, Infrastructure, and Geopolitics Project," Chicago Council on Global Affairs, https://globalaffairs.org/research/center-global-cities/cities-infrastructure-and-geopolitics-project.

25. Jonathan Rokem and Camillo Boano, eds., *Urban Geopolitics: Rethinking Planning in Contested Cities* (London: Routledge, 2018); Voula P. Mega, *Conscious Coastal Cities: Sustainability, Blue Green Growth, and the Politics of Imagination* (Cham: Springer, 2016), 1–38.

26. Stephen Graham and Sárka Waisová, "Cities, War, and Terrorism: Towards an Urban Geopolitics," *Journal of International Relations and Development* 10, no. 1 (2007): 90–92; Colin McFarlane and Jennifer Robinson, "Introduction-Experiments in Comparative Urbanism," *Urban Geography* 33, no. 6 (2012): 765–73; Jonathan Rokem et al., "Interventions in Urban Geopolitics," *Political Geography* 61 (2017): 253–62; Jonathan Rokem and Camillo Boano, "Towards a Global Urban Geopolitics: Inhabiting Violence," *Geopolitics* 28, no. 5 (2023): 1667–80.

27. Rokem and Boano, *Urban Geopolitics*.
28. "Our Future City: The Atlanta City Design," City of Atlanta, GA, https://www.atlantaga. gov/home/showdocument?id=30594&ref=welcometohellworld.com.
29. "The Center for Diversion & Services," Policing Alternatives & Diversion Initiative, https://www.atlantapad.org/diversion-center.
30. "A Timeline of All Atlanta's Past Mayors," WABE, October 25, 2017, https://www.wabe. org/timeline-atlantas-past-mayors/.
31. Alan Blinder and Nicole Perlroth, "A Cyberattack Hobbles Atlanta, and Security Experts Shudder," *New York Times*, March 27, 2018, https://www.nytimes.com/2018/03/27/us/ cyberattack-atlanta-ransomware.html.
32. Richard Fausset, "'Covid Crime Wave' Weighed Heavily on Atlanta Mayor," *New York Times*, May 7, 2021, https://www.nytimes.com/2021/05/07/us/covid-crime-keisha-lance-bottoms.html.
33. "Text: Atlanta Mayor's Speech to City during Violent Protests," 11Alive, May 30, 2020, https://www.11alive.com/article/news/local/atlanta-protests-mayor-speech-full-text/85-865bb430-7502-44fa-a9b4-3414332ec342.
34. Achille Mbembé, "Necropolitics," trans. Libby Meintjes, *Public Culture* 15, no. 1 (2003): 11.
35. Juan M. Floyd-Thomas, "'A Relatively New Discovery in the Modern West': #BlackLivesMatter and the Evolution of Black Humanism," *Kalfou* 4, no. 1 (2017): 30–39; Taylor, *From #BlackLivesMatter to Black Liberation*, 191–219; Lawrence T. Brown, "The Movement for Black Lives vs. the Black Church," *Kalfou* 4, no. 1 (2017): 7–17.
36. Philip L. Reichel, "Southern Slave Patrols as a Transitional Police Type," *American Journal of Police* 7, no. 2 (1988): 51–77; Pamela Collins et al., *Principles of Security and Crime Prevention*, 4th ed. (New York: Routledge, 2015); and K. B. Turner, David Giacopassi, and Margaret Vandiver. "Ignoring the Past: Coverage of Slavery and Slave Patrols in Criminal Justice Texts," *Journal of Criminal Justice Education* 17, no. 1 (2006): 181–95.
37. Ruth Wilson Gilmore, *Golden Gulag: Prisons, Surplus, Crisis, and Opposition in Globalizing California* (Berkeley: University of California Press, 2007), 28.
38. Sandra Browne, *Dark Matters: On the Surveillance of Blackness* (Durham, NC: Duke University Press, 2015).
39. Richard Fausset, "Andre Dickens, a Veteran City Council Member, Is Elected Mayor of Atlanta," *New York Times*, November 30, 2021, https://www.nytimes.com/2021/11/30/ us/andre-dickens-atlanta-mayor-election.html.
40. "Moving Atlanta Forward Agenda," City of Atlanta, GA, https://www.atlantaga.gov/ government/mayor-s-office/moving-atlanta-forward-agenda.
41. "Meet the Mayor," City of Atlanta, GA, accessed September 25, 2023, https://www.

atlantaga.gov/government/mayor-s-office/meet-the-mayor.

42. Adam Gabbatt, "Stark Warning over Republicans' 'Dehumanizing' Rhetoric on Crime," *The Guardian*, May 14, 2023, https://www.theguardian.com/us-news/2023/may/14/republican-tough-on-crime-us-elections2010.

43. Ames Grawert and Noah Kim, "Myths and Realities: Understanding Recent Trends in Violent Crime" Brennan Center for Justice, July 12, 2022, https://www.brennancenter.org/our-work/research-reports/myths-and-realities-understanding-recent-trends-violent-crime.

44. "Homicide, Gun Assault, Domestic Violence Declined in Major U.S. Cities in 2022 but Remain above Pre-Pandemic Levels," Council on Criminal Justice, January 26, 2023, https://counciloncj.org/homicide-gun-assault-domestic-violence-declined-in-major-u-s-cities-in-2022-but-remain-above-pre-pandemic-levels/.

45. The U.S. Crisis Monitor, https://acleddata.com/special-projects/us-crisis-monitor/, a joint project between the Armed Conflict Location & Event Data Project and the Bridging Divides Initiative at Princeton University, collects real-time data on these trends to provide timely analysis and resources to support civil society efforts to track, prevent, and mitigate the risk of political violence in America. With supplemental data collection extending coverage back to the week of Floyd's killing in late May 2020, the dataset now encompasses the latest phase of the Black Lives Matter movement, growing unrest related to the health crisis, and politically motivated violence ahead of the November 2020 general election.

46. Floyd-Thomas, "'A Relatively New Discovery in the Modern West'"; Christopher Cameron and Phillip Luke Sinitiere, eds., *Race, Religion, and Black Lives Matter: Essays on a Moment and a Movement* (Nashville: Vanderbilt University Press, 2021); and Christophe D. Ringer, Teresa L. Smallwood, and Emilie M. Townes, eds., *Moved by the Spirit: Religion and the Movement for Black Lives* (Lanham, MD: Lexington Books, 2023).

47. Kerby Goff and John D. McCarthy, "Critics Claim BLM Protests Were More Violent than 1960s Civil Rights Ones: That's Just Not True," *Washington Post*, October 12, 2021, https://www.washingtonpost.com/politics/2021/10/12/critics-claim-blm-was-more-violent-than-1960s-civil-rights-protests-thats-just-not-true/.

48. Timothy Pratt, "Activists Push for Referendum to Put 'Cop City' on Ballot in Atlanta," *The Guardian*, June 16, 2023, https://www.theguardian.com/us-news/2023/jun/16/atlanta-cop-city-activists-push-referendum?utm_term=648c4ec3bacab394fb26f53ab4b48580&utm_campaign=GuardianTodayUS&utm_source=esp&utm_medium=Email&CMP=GTUS_email.

49. Ta-Nehisi Coates, "The Case for Reparations," *The Atlantic*, June 2014, https://www.

theatlantic.com/magazine/archive/2014/06/the-case-for-reparations/361631/; Katherine McKittrick, *Demonic Grounds: Black Women and the Cartographies of Struggle* (Minneapolis: University of Minnesota Press, 2006); Rothstein, *The Color of Law*; Meghan Cope and Frank Latcham, "Narratives of Decline: Race, Poverty, and Youth in the Context of Postindustrial Urban Angst," *Professional Geographer* 61, no. 2 (2009): 150–63; Joe T. Darden and Elvin Wyly, "Cartographic Editorial—Mapping the Racial/ Ethnic Topography of Subprime Inequality in Urban America," *Urban Geography* 31, no. 4 (2010): 425–33; Mustafa Dikeç, "Badlands of the Republic? Revolts, the French State, and the Question of Banlieues," *Environment and Planning D: Society and Space* 24, no. 2 (April 2006): 159–63; Mary Pattillo, "Extending the Boundaries and Definition of the Ghetto," *Ethnic and Racial Studies* 26, no. 6 (2003): 1046–57; Rashad Shabazz, *Spatializing Blackness: Architectures of Confinement and Black Masculinity in Chicago* (Champaign: University of Illinois Press, 2015); David Wilson, *Cities and Race: America's New Black Ghetto* (London: Routledge, 2007); David Wilson and Carolina Sternberg, "Changing Realities: The New Racialized Redevelopment Rhetoric in Chicago," *Urban Geography* 33, no. 7 (2012): 979–99; William J. Wilson, *The Truly Disadvantaged: The Inner City, the Underclass, and Public Policy* (Chicago: University of Chicago Press, 1987); Taylor, *Race for Profit*; Elvin Wyly et al., "Cartographies of Race and Class: Mapping the Class-Monopoly Rents of American Subprime Mortgage Capital," *International Journal of Urban and Regional Research* 33, no. 2 (2009): 332–54; Robert Blauner, "Internal Colonialism and Ghetto Revolt," *Social Problems* 16, no. 4 (1969): 393–408.

50. Keeanga-Yamahtta Taylor, *From #BlackLivesMatter to Black Liberation* (Chicago: Haymarket Books, 2016).

51. Deborah Cowen and Nemoy Lewis, "Anti-Blackness and Urban Geopolitical Economy," Society+Space, August 2, 2016, https://www.societyandspace.org/articles/anti-blackness-and-urban-geopolitical-economy.

52. Juan M. Floyd-Thomas, "The Donald Went Down to Georgia: The GOP from God's Own Party to the Party of Trump," in *Faith and Reckoning after Trump*, ed. Miguel A. De La Torre (Maryknoll, NY: Orbis Books, 2021), 13–24.

53. Jael Holzman, "Georgia Senators Criticize Damage at 'Cop City' Protests," AXIOS online, March 9, 2023, https://www.axios.com/local/atlanta/2023/03/09/georgia-senators-criticize-damage-at-cop-city-protests.

54. Patrick Quinn, "'Stop Cop City' Sen. Warnock Interrupted during Commencement Address," Atlanta News First, May 29, 2023, https://www.atlantanewsfirst.com/2023/05/29/stop-cop-city-sen-warnock-interrupted-during-commencement-address/?outputType=amp.

55. Tia Mitchell, "Warnock, Black Caucus Push Biden to Make Policing a Theme of State of the Union," *Atlanta Journal-Constitution*, February 6, 2023, https://www.ajc.com/politics/warnock-black-caucus-push-biden-to-make-policing-a-theme-of-state-of-the-union/USLEXTSNFJHVJCETMG7CFWFQ7Y/.

56. Ryan Zickgraf, "The Terrorism Charges against Cop City Protesters Are Ominous," *Jacobin*, January 27, 2023, https://jacobin.com/2023/01/domestic-terrorism-charges-cop-city-protests-atlanta-georgia.

57. "Text—H.Res.252—Condemning the Violent 'Stop Cop City' Movement in Atlanta, Georgia," 118th Congress (2023–2024), Congress.gov, March 24, 2023, https://www.congress.gov/bill/118th-congress/house-resolution/252/text.

58. "Sen. Ossoff Secures Answers from FBI Director on Causes of Increased Violence in Atlanta," Jon Ossoff, U.S. Senator for Georgia, September 21, 2021, https://www.ossoff.senate.gov/press-releases/sen-ossoff-secures-answers-from-fbi-director-on-causes-of-increased-violence-in-atlanta/.

59. Taiyler S. Mitchell, "Dem Lawmakers Speak Out against Cop City Arrests," HuffPost, June 5, 2023, https://www.huffpost.com/entry/lawmakers-cop-city-arrests-atlanta_n_647cf7a3e4b02325c5e1608e/amp.

60. Sharon Zhang, "Warnock Probes Whether State Trampled Free Speech Rights of Cop City Protesters," Truthout, June 8, 2023, https://truthout.org/articles/warnock-government-may-be-blocking-free-speech-rights-of-cop-city-protesters/.

61. "Press Release—Mayor Andre Dickens Joins Senator Jon Ossoff to Announce $30 Million in Safety Upgrades for Pedestrians in Downtown Atlanta," City of Atlanta, GA, February 7, 2023, https://www.atlantaga.gov/Home/Components/News/News/14557/672?backlist=%2F.

62. Baldwin and Clover, "Police Academies Are Part of Higher Ed."

63. John Sloan, "Police Academies Even More Associated with Higher Ed than Essay Detailed," *Chronicle of Higher Education*, August 24, 2023, https://www.chronicle.com/blogs/letters/police-academies-even-more-associated-with-higher-ed-than-essay-detailed.

64. Caitlin Lynch, "You Have the Right to Remain Violent: Police Academy Curricula and the Facilitation of Police Overreach," *Social Justice* 45, no. 2–3 (2019): 75–92.

65. "Atlanta Committee for Progress to Support Mayor Bottoms' Plan to Address Violent Crime," Atlanta Committee for Progress, April 1, 2021, https://atlprogress.org/_pdf/ACP_Public_Safety_Release_04-01-21.pdf.

66. Taya Graham and Stephen Janis, "Atlanta's 'Cop City' Is a Blueprint for America's Future," Real News Network, April 3, 2023, https://therealnews.com/atlantas-cop-city-is-a-blueprint-for-americas-future.

67. "Cop City: Atlanta City Council OKs $67M for Facility despite Mass Protests & Armed Raid on Bail Fund," Democracy Now!, June 7, 2023, https://www.democracynow.org/2023/6/6/cop_city_atlanta_vote_bail_fund.

68. George Chidi, "No One Believes in Cop City: So Why Did Atlanta's City Council Fund It?," *The Intercept*, June 7, 2023, https://theintercept.com/2023/06/06/cop-city-atlanta-funding-vote/.

69. Micah Herskind, "This Is the Atlanta Way: A Primer on Cop City," *Scalawag*, May 2, 2023, https://scalawagmagazine.org/2023/05/cop-city-atlanta-history-timeline/.

70. Gilmore, *Golden Gulag*, 26.

71. Herskind, "This Is the Atlanta Way."

Part 3

Decolonizing and Implementing Environmental Justice

Material Flows in Landscapes of Injustice

Nikiwe Solomon

C ovid-19," an informal settlement, aptly named for being established during the pandemic lockdown in the year 2020, is located in the Driftsands Nature Reserve with the Kuils River flowing through it in the Cape Town metropolitan area in South Africa. The settlement composed of informal shack dwellings is home to thousands, with reported numbers ranging from eleven thousand to twenty-seven thousand people settling on what the City of Cape Town has argued is private land.[1] According to the Cape Town Green Map website, Driftsands is "a provincial nature reserve run by CapeNature Conservation" that provides "environmental education activities and is strategically located in an area where it can provide access to nature for impoverished communities. Driftsands also protects remnants of endangered vegetation types Cape Flats Dune Strandveld."[2]

In 2016, I visited the reserve and met with the manager who was rather hesitant to speak to me; in hindsight, this was due to the complexities of managing a space fraught with various forms of conflict including whether the protection of plants and species should supersede people's access to land for settlement, especially in a part of the city that has significant levels of poverty and violence.

The Driftsands is a nine hundred hectare reserve located in one of the most densely populated areas of Cape Town characterized by informal and low-cost housing, with the population facing significant socioeconomic and health challenges. As such, during my visit to the reserve, the discussion of the role of a nature reserve located in such a landscape, on the margins of a city marred by high levels of inequality, ensued. The plan was to provide a "multi-functional urban park, combining biodiversity conservation with job creation and local economic development" (capenature.co.za).[3] However, according to the nature reserve manager, the biggest threat to the conservation efforts, particularly to the endangered Strandveld fynbos were so-called land invasions, brought on by a rapidly expanding urban population. A few years later, the perceived threat is now a reality with the establishment of the "Covid-19" settlement within the reserve.

This settlement, one of many that sprung up during the pandemic lockdown, has been a site of contention and different forms of violence since its inception. Eviction notices were served in June 2022 resulting in protests, which led to a clash between residents and the police with rubber bullets fired at protestors who blocked the entrance to the nature reserve. In February 2023, four people were shot and killed under unknown circumstances, with residents fearing to go out and help because of the lack of streetlights in the area. In June 2023, the residents feared disease outbreak as the settlement is located on seasonal wetlands that are inundated during Cape Town's winters, which are the peak rainfall periods in the year. As the area has no formal sanitation services, residents feared that human waste disposed of in the vicinity would enter their homes, posing serious health risks. In September 2023, four people were electrocuted and died as homes had illegal electricity connections and were flooded during a storm when the banks of the Kuils River were breached.

So-called land invasions in South Africa are commonplace; however, according to those in power and the mainstream press, since the coronavirus outbreak, there had been a staggering rise in "illegal" land occupations, while ignoring the long history of land dispossession under colonialism and apartheid. Colonialism and apartheid shaped the significant economic and geographical divisions in contemporary South Africa. Those who lack the financial means to secure accommodation will find a place to live by any means necessary, particularly in urban areas where one is likely to find paid work. In Cape Town, under the COVID-19 lockdown restrictions, many residents lost their jobs and income, meaning they could no longer afford to pay their rent; therefore spaces like the Covid-19 informal settlement, and another called "Sanitizer" next to it (an example of South Africans' sense of humor even in tough times), emerged across the city where they could live rent free.

While the Covid-19 settlement exemplifies the statement COVID-19 has made visible the fault lines of inequality in society, for many residents across the globe, these fault lines were always visible in the toxic landscapes in which they have lived. For instance, in my PhD research, I refer to residents living a few kilometers upstream near the Kuils River having complained about respiratory problems since the reopening of a steel processing plant owned by DHT Holding, trading as Cisco. The plant originally operated from 1960 to 2010 and was shut down until it was bought by DHT Holding in 2012. In this period, the Kuils River was largely farmland and had a significant buffer area between the steel mill and residential areas. However, when the mill shut down, housing developments sprung up in the area, and the buffer area between the mill and residential area had decreased to about seventy meters when the mill reopened in 2018. While the health of residents living in the area is cause for concern, city officials and the owner of the mill are hesitant in linking the growing respiratory issues to the mill, as there is "no proof" of a direct link to the plant's emissions. DHT Holding argues that a significant 550 million rand from the company and an additional 230 million rand from the Department of Trade and Industry went into upgrading the facility, which included improving emissions from the plant. Moreover, the plant provides three hundred jobs, which can ultimately improve the lives of locals. In 2019, the City of Cape Town's own air quality officer on the other hand stated that the lack of a buffer zone between the steel mill and the Kuils River residential area was a "land-use planning disaster."[4]

Not only were residents concerned about the air and noise pollution from the mill, a study conducted by Shezi et al. in preschool facilities around industrial operations in the Kuils River area showed that contamination in the soil samples collected from the gardens of some of the study areas existed. For Shezi et al., "The health index (HI) for non-carcinogenic effects showed the ingestion route as the main contributor to the total risk, with cumulative carcinogenic risk exceeding the maximum acceptable level."[5] At some of the preschools, arsenic and zinc were found to exceed Canadian soil reference levels (but below South African reference levels). However, because of the types of soils characteristic in the area, which are sandy and often acidic, "heavy metals are more soluble in acids, and therefore acidification may increase bioaccumulation of heavy metals and result in increased exposure."[6] The city's response was that they had noted the study, highlighting that "there are a number of historic sources that could have led to the findings; . . . the study did not identify CISCO as the source of all the pollutants." They also stated that they "head [sic] the study recommendations when it comes to authorising further listed activities in the area that could impact of [sic] heavy metals surface contamination of the Kuils River area,"[7] reinforcing their stance from previous years that there is no proof of a direct link between contamination of the area and the steel mill.

If there is already reluctance to conclude that contamination of the soil could be from the mill, justifying an investigation into how pollution has affected different bodies (water, multiple species to name a few), transporting sediments in water, air, dust, mud, downstream, and into the ocean, would prove even more difficult than proving harm to humans. In addition, it would be even more difficult to prove that the area of Kuils River that had some of the highest COVID infection rates in the city were because of their proximity to polluting industries.[8]

The focus has been rather to shift the blame to individual behavior. And yet, a study by Cole et al. shows that there is compelling evidence of a positive relationship between air pollution and COVID-19 cases, hospital admissions, and death.[9] Studies have shown that long-term exposure to air pollution was related to increased hospitalization, experience of severe COVID-19, and likelihood of death as air pollution is known to increase the numbers or the intensity of cardiovascular and respiratory disease, which are known to increase the susceptibility to the virus.[10] Other studies conducted during the SARS outbreak showed that increased air pollution could be linked to poorer outcomes of infection such as hospitalization and death.[11]

Rupa and Patel argue that constant exposure to toxic environments sometimes results in a chronic systemic inflammatory response, where the body's immune system breaks down, diminishing its ability to "repair damage and restore homoeostasis," a state in which the body operates at its optimum levels.[12] Constant inflammation of the body due to nongenetic drivers of health and illness throughout a lifetime, called the exposome, can include "chemical, social, psychological, ecological, historical, political, and biological elements that determine whether aging cells will become drivers of chronic systemic inflammation."[13] When the immune system is shaped by the exposome, its reaction to a virus like COVID pushes it into an overdrive of inflammation. The authors argue that severe COVID is expressed in socially oppressed groups, where the socioeconomic and environmental exposome impact is immediate but also changeable. But for those who live these precarious lives, the exposome is not optional; it is systematic.[14]

The response from the city and operators of the steel mill raises the question: Does a corporate entity's right to operate and conduct business outweigh citizens' rights to a clean environment and good health? No precautionary measures were taken to ensure the health and safety of the human and multispecies communities in the areas where toxic sediments could be transported through the air, water, and other vectors, thereby exposing them to slow forms of violence and a health risk that could lead to premature death.[15]

Steve Lerner proposes a compelling argument for why prevailing environmental management must be reexamined to place emphasis on the precautionary principle, prevention, and commensurate protection.[16] This argument, the result of two years of research and work with twelve communities in the middle of toxic "sacrifice zones" in the United States, is backed by irrefutable evidence that not all Americans are created equal.[17] Lerner's book, *Sacrifice Zones*, reveals that one of the most important indicators of an individual's health is their ZIP code, which corresponds to histories of segregation between communities of color and white communities.[18] For Lerner, this pattern of unequal protections constitutes environmental racism, as these spaces are often occupied by low-income people of color,[19] a trend we also see in the Cape Town context. In such spaces, the well-being of people and the environment are sidelined in the name of "economic development" and "progress," often brought about by technical proposals and responses assumed to be objective and neutral.[20]

Zones of Sacrifice and Slow Violence

In both cases, in the Covid-19 settlement and the Kuils River neighborhood, time and place matter. Rupa and Patel highlight that "experiencing daily trauma at the hands of law enforcement, acute hunger, discrimination, forced displacement, and disproportionate exposure to toxins—it all makes people sick. If every diagnosis is a story, and every story begins with 'once upon a time in a faraway place,' the specific time and place matter."[21] The cases of the Covid-19 settlement and the Kuils River neighborhood provide insight into experiences of living with the polluted landscapes and what some call "fortress conservation,"[22] its impacts on people's health, and the difficulties they have in proving that harm in the age of the

Anthropocene. While residents of the area expressed their concerns, by speaking to reporters, protesting, and many different avenues, the response was fixated on the science of numbers and what the numbers were assumed to portray. Because these numbers are seen as fact and drawn from scientific fact, they often do not take into account the actual experiences of living with the pollutants represented by these numbers. Listening to experiences of people is often not recognized or considered within the realms of scientific knowledge production, which is more focused on data.

The Anthropocene, the era in which humans have significantly altered the Earth's spheres, is often seen as occupying two different realms: one of science and measurement that calls for pragmatic objective enquiry, and the other of the mind and body that elicits moral responses and "behaviour change."[23] If these realms remain separate, how we think of geological time in the social sciences could render the environment and geological relations as a space of dormancy until humans came along. Yet as evidenced by many geological periods over millions of years, the environment is above all else a temporal feature in constant flux, and it is never the same from moment to moment. When Cape Town's geology, environmental connections, and flows are rendered to the background of the human experience, what Aime Cesaire calls "thingified,"[24] the Driftsands Nature Reserve and the soils and air of the Kuils River neighborhood are regarded simply as space, and the geochemical processes that make these landscapes possible get no attention. The flows and ebbs that make lives and worlds possible or impossible are invisibilized, and using the same logics that justified colonialism, the kinds of harm enacted on it are justified because the spaces are "empty" spaces of no significance.

Environmental health and human health have been shown by researchers to be inextricably linked,[25] but the complexities of providing evidence of environmental harm to a specific standard of both evidence and direct causality mean that such an association is easily disregarded. Strategies for environmental management often overlook knowledge claims by the communities that live in toxic spaces or the research findings of scientists that focus on the impact of pollution on people and multispecies communities. This makes certain populations and geographies vulnerable to environmental sacrifice.[26]

Rob Nixon interprets this type of interaction between time and environmental degradation using the concept of slow violence. Violence is often understood as spectacular and immediate, but Nixon reminds us that it need not be to have damaging consequences.[27] Slow violence is often a result of uneven social conditions, largely through the colonial project of displacement and dispossession. Those most impacted are often the marginalized members of society, settled in hazardous spaces through discriminatory geological categorizations.

Sacrifice zones are often located in edge-dwelling communities made up primarily of people of color with low incomes, or are hot spots of chemical pollution, where residents live in close proximity to industrial areas or heavily polluted streams (such as the Kuils).[28] In South Africa, these areas also lack adequate housing, water, and sanitation infrastructure. As Davies suggests, these patterns of unequal services and protections often constitute environmental racism, demonstrating that environmental justice is not just a "poverty thing" but is a result of systemic and institutional patterns of inequality that are swept under the rug of democracy.[29] I argue that how this is enacted is a form of violence on human and multispecies communities, some explicit and visible and others as forms of slow violence, which is cumulative and has potentially cascading effects. Rob Nixon refers to slow violence as a violence that

> occurs gradually and out of sight, an attritional violence that is typically not viewed as violence at all. Violence is customarily conceived as an event or action that is immediate in time, explosive and spectacular in space, and as erupting into instant sensational visibility. We need, I believe, to engage a different kind of violence, a violence that is neither spectacular nor instantaneous, but rather incremental and accretive, its calamitous repercussions playing out across a range of temporal scales.[30]

For whom is this violence gradual and out of sight, Davies asks.[31]

The Kuils River community has been intensely frustrated, because not enough was done with any urgency to alleviate the problems of pollution as this would affect production cycles and profitability margins of industries in the area. Therefore, while the operation of industries is deemed necessary to societal goals of development (more often than not determined and imposed by society's elite), they have often resulted in the creation of sacrifice zones where lives (human, flora, and fauna) are regarded as cheap and disposable in the interests of economic and political opportunity.

City planning in Cape Town was historically designed to provide services to an elite white and wealthy minority through infrastructure such as roads, water pipes, sewage removal pipes, and power lines; the city offered limited services to people of color. The neighborhoods for people of color were also placed in landscapes in close proximity to industries, not only for the supply of labor, but also because these spaces were seen as unproductive under colonial and apartheid regimes. The disparity in services provided and environmental protections offered to wealthy (mostly white) residents and low-income (mostly colored and Black) residents produces different versions of urban living. The racial necropolitics of settlement patterns in the city shows landscapes along the Kuils River are embedded with differentiated experiences, resulting in covert and difficult-to-prove cases of harm. Limiting the epistemics of harm to proof of direct causality, as required by the rhetorical statements of City of Cape Town political leaders and officials, enacts a form of necropolitics. Mbembe describes necropolitics as the ultimate expression of power by a sovereign (or state), deciding who thrives and who does not.[32] By exploring the necropolitics of the Kuils River landscapes, we can see how exposure to chemical pollutants, poor and inadequate infrastructure, and "fortress conservation" along the river create harmful spaces that are a perpetual threat to the well-being of people and the environment. These threats are experienced in covert and limiting ways, making it difficult to prove harm and giving the state the power to determine what should be deemed life-threatening and what should not.

Mbembe's argument of necropolitics, an extension of Michel Foucault's notion of biopower,[33] which expresses the power to control, manage, and determine life, is useful in understanding the dynamics at play for the Covid-19 settlement and the Kuils River community. In these cases, biopower and necropolitics are manifest in the centuries of displacement of people of color and the ongoing destruction of the environment, exposing communities (human and other-than-human) to hazardous landscapes through covert forms of violence. The violence enacted on such populations and multispecies communities is considered slow and subtle, often gaslighting its victims though a lack of catastrophic or spectacular evidence to demonstrate injury. Under the guise of economic development, the rhetoric of conservation, and urban renewal programs (and by extension citizens' lives), the paradoxical nature of the destruction of other lives and systems is not seen or is ignored.

Mbembe argues that such violence does not originate from a single event or power but is entangled in a complex assemblage of capitalist structures, government authority, and corporate and legislative power.[34] In such cases, environmental harm is often not intentional, but the result of environmental management and of scientific, governance, and neoliberalism machinations.[35]

This is often informed by how society and the environment are seen as separate, with material flows between the two invisibilized, ignored, or just undermined. What happens when the environment is seen as just space, something out there? It becomes easy to place borders, fences, boundaries enforced through policing or fines, criminalizing already vulnerable groups. It is easier to dismiss claims of harm when the wind blows dangerous dust into preschools and homes, or when water moves toxic sediments down the river. As such, the concept of urban metabolism becomes key to understanding human relationships to the environment.[36] Urban metabolism refers to the material and energy flows in cities that are shaped by social, economic, and environmental forces to create a complex system. Clarke and Foster's adaptation of Karl Marx's notion of the metabolic rift is useful for thinking about this imposed separation between people and the environment under modernity.[37] For Clarke and Foster, the metabolic rift is a result of the alienation of people from the Earth and Earth cycles, and this can be used to explain the rise in global ecological crises.[38]

Governance of Space as Governance of Territory or Flow?

Urban environmental management in the city is dominated by scientific and engineering approaches and solutions to environmental problems that emanate from what are claimed to be "objective" standpoints. However, these so-called objective approaches are not neutral, but are a result of a social, political, and cultural imagination of the environment as just space from which we derive "ecosystem services." When the environment is presented as a singular unified object that exists in a landscape out there in "nature" for extraction by humans, the landscapes are presented as "just space," taking it out of its relationships to people and multispecies communities.[39] Drawing on one perspective of what these landscapes are and what their role in the city is imagined to be makes invisible the complex urban ecology in which service delivery protests are many, infrastructure is failing, wetlands and biodiversity are declining, the climate is changing, and new forms of chemicals are entering into the environment. Understanding these connections—the urban flows of air, water, chemicals, nutrients, and energy entangled with our own bodies—is therefore important if we are to respond carefully and justly in the time of the Anthropocene.

The call is therefore for a regime shift from governance of landscapes as territory, bound in space and time, to a governance that centers flows not only through landscapes, but also through time. Shifting the concern from how to control, manage, and predict human entanglements with the environment, to concern of things flowing through geographies, bodies, infrastructures, politics, economies, and time, places well-being of people and environments as a priority. Bodies are inflamed when they are exposed to a constant barrage of threats, so how we take care and repair matters. COVID has shown us why this matters. For Rupa and Patel, to center care in a time of climate crisis and pandemics requires a reframing of liberal questions of property of "Who gets what?" into "Who does what?" alongside "Who has had what done to them?"[40] These concerns alongside understanding relationships and flows begin the work of repairing injuries resulting from hundreds of years of harm caused by colonial encounter. Healing and repair require a reconfiguration of cosmologies that allow for the intricate links between humans and the environment to set center stage, and not to exist as separate.

NOTES

Parts of the work in this paper were drawn from my PhD thesis, titled "The Kuils River Multiple: Versions of an Urban River on the edge of Cape Town, South Africa," which was made possible with support from the National Institute for Humanities and Social Sciences (NIHSS). The financial assistance of the NIHSS, in collaboration with the South African Humanities Deans Association (SAHUDA), toward this research is hereby acknowledged. Opinions expressed and conclusions arrived at are those of the author and are not necessarily to be attributed to the NIHSS and SAHUDA.

1. See Thina Nzo, "Covid-19 Informal Settlement Delegitimizes Authority of the City of Cape Town," *Daily Maverick*, June 29, 2022, https://www.dailymaverick.co.za/opinionista/2022-06-29-covid-19-informal-settlement-delegitimises-authority-of-the-city-of-cape-town/; also see Philani Nombembe, "Now Covid-19 Kills Nature Reserve as Invaders Unleash Rubble Trouble," *Sunday Times*, May 8, 2022, https://www.timeslive.co.za/sunday-times/news/2022-05-08-now-covid-19-kills-nature-reserve-as-invaders-unleash-rubble-trouble/.

2. "Driftsands Nature Reserve," Cape Town Green Map, https://www.capetowngreenmap.

co.za/cape-town-green-map-online-map/nature/park-recreation-area/driftsands-nature-reserve.

3. L. Saul, N. Hayward, G. Cleaver-Christie, and T. Maliehe, "Driftsands Nature Reserve, Western Cape, South Africa: Protected Area Management Plan 2015–2020," Cape Nature, March 2015, https://www.capenature.co.za/uploads/files/protected-area-management-plans/DRFS-PAMPS_merged-1.pdf.

4. Tred Magill, "Cape Town Residents Fear Threat of Illness, Air Pollution as Steel Mill Starts up too Close to Homes," *News24*, March 13, 2023, https://www.news24.com/news24/southafrica/news/cape-town-community-fears-threat-of-illness-air-pollution-as-steel-mill-start-up-too-close-to-homes-20230313.

5. Busisiwe Shezi et al., "Heavy Metal Contamination of Soil in Preschool Facilities around Industrial Operations, Kuils River, Cape Town (South Africa)," *International Journal of Environmental Research and Public Health* 19, no. 4380 (2022): 1, https://www.mdpi.com/1660-4601/19/7/4380.

6. Shezi et al., "Heavy Metal Contamination," 10.

7. Activist email to author. The activist's name is kept anonymous for ethical reasons.

8. See Shakirah Thebus, "Resurgence of Covid-19 Is Spreading like Wildfire in the Western Cape," *IOL News*, November 27, 2020, https://www.iol.co.za/capeargus/news/resurgence-of-covid-19-is-spreading-like-wildfire-in-the-western-cape-fe910044-7cc1-488d-a8af-39484bd40e70.

9. Matthew A. Cole, Ceren Ozgen, and Eric Strobl, "Air Pollution Exposure and Covid-19 in Dutch Municipalities," *Environmental and Resource Economics* 76, no. 4 (2020): 581–610, doi: 10.1007/s10640-020-00491-4.

10. Jiawei Zhang et al., "Long-Term Exposure to Air Pollution and Risk of SARS-CoV-2 Infection and COVID-19 Hospitalization or Death: Danish Nationwide Cohort Study," *European Respiratory Journal* 62, no. 1 (2023), doi: 10.1183/13993003.00280-2023. PMID: 37343976; PMCID: PMC10288813; Ireri Hernandez Carballo, Maria Bakola, and David Stuckler, "The Impact of Air Pollution on COVID-19 Incidence, Severity, and Mortality: A Systemic Review of Studies in Europe and North America," *Environmental Research* 215 (2022): 1–14, doi: 10.1016/j.envres.2022.114155.

11. Yan Cui et al., "Air Pollution and Case Fatality of SARS in the People's Republic of China: an Ecologic Study," *Environmental Health* 2, no. 15 (2003), https://doi.org/10.1186/1476-069X-2-15.

12. Marya Rupa and Raj Patel, *Inflamed: Deep Medicine and the Anatomy of Injustice* (London: Penguin, 2021), 30.

13. Rupa and Patel, *Inflamed*, 32.

14. Rupa and Patel, *Inflamed*, 34.

15. Robert Nixon, *Slow Violence and the Environmentalism of the Poor* (Cambridge, MA: Harvard University Press, 2011), 2; Achille Mbembe, *Necropolitics* (Durham, NC: Duke University Press, 2019).

16. Steve Lerner, *Sacrifice Zones: The Front Lines of Toxic Chemical Exposure in the United States* (Cambridge, MA: MIT Press, 2010).

17. Robert D. Bullard, review of *Sacrifice Zones: The Front Lines of Toxic Chemical Exposure* by Steve Lerner, *Environmental Health Perspectives* 119, no. 6 (2011): A266, https://www.ncbi.nlm.nih.gov/pmc/articles/PMC3114843/.

18. Lerner, *Sacrifice Zones*, 18.

19. In Bullard, review of *Sacrifice Zones*, 2.

20. Nikiwe Solomon, "The Kuils River Multiple: Versions of an Urban River on the Edge of Cape Town South Africa" (PhD diss., University of Cape Town, 2021), 14.

21. Rupa and Patel, *Inflamed*, 9.

22. See David Hill, "Rights, Not 'Fortress Conservation,' Key to Saving Planet, Says UN Expert," *The Guardian*, July 17, 2018, https://www.theguardian.com/environment/andes-to-the-amazon/2018/jul/16/rights-not-fortress-conservation-key-to-save-planet-says-un-expert; Irene Wabiwa Betoko and Savio Carvalho, "To Protect Nature, Bring down the Walls of Fortress Conservation," Greenpeace, October 20, 2020, https://www.greenpeace.org/international/story/45497/indigenous-people-biodiversity-fortress-conservation-power-shift/.

23. Dipesh Chakrabarty, "Anthropocene Time," *History and Theory* 57, no. 1 (2018): 6–32, https://doi.org/10.1111/hith.12044

24. Aime Cesaire, *Discourse on Colonialism*, trans. Joan Pinkham (New York: Monthly Review Press, 2000).

25. See Shani Fourie, "An Assessment of Water Quality and Endocrine Disruption Activities in the Eerste/Kuils River Catchment System, Western Cape, South Africa" (master's thesis, Stellenbosch University, 2005); also see Jeanne Nel et al., "South Africa's Strategic Water Source Areas: Report for WWF-SA" (report no. CSIR/NRE/ECOS/ER/2013/0031/A, 2013), and François Ngera Mwangi, "Land Use Practices and Their Impact on Water Quality of the Upper Kuils River, Western Cape Province, South Africa" (master's thesis, University of Western Cape, 2014).

26. Thom Davies, "Slow Violence and Toxic Geographies: 'Out of Sight' to Whom?," *Environment and Planning C: Politics and Space* 40, no. 2 (2022), https://doi.org/10.1177/2399654419841063.

27. Nixon, *Slow Violence*, 6.

28. Thom Davies, "Toxic Space and Time: Slow Violence, Necropolitics, and Petrochemical Pollution," *Annals of the American Association of Geographers* 108, no. 6 (2018): 1537–53, https://doi.org/10.1080/24694452.2018.1470924.

29. Mbembe, *Necropolitics*.

30. Nixon, *Slow Violence*, 2.

31. Davies, "Slow Violence and Toxic Geographies."

32. Mbembe, *Necropolitics*, 66.

33. Michel Foucault, *The Birth of Biopolitics, Lectures at College de France 1977–1979* (New York: Picador, 2010).

34. Mbembe, *Necropolitics*, 66.

35. Davies, "Toxic Space and Time."

36. See Giles Thomson and Peter Newman, "Urban Fabrics and Urban Metabolism—From Sustainable to Regenerative Cities," *Resources, Conservation and Recycling*, no. 132 (2018): 218–29. Also see Teresa Laginha Sanches and Nuno Ventura Santos Bento, "Urban Metabolism: A Tool to Accelerate the Transition to a Circular Economy," in *Sustainable Cities and Communities: Encyclopedia of the UN Sustainable Development Goals*, ed. Walter Leal Filho, Anabela Marisa Azul, Luciana Brandli, Pinar Gökçin Özuyar, and Tony Wall (Cham: Springer, 2020), https://doi.org/10.1007/978-3-319-95717-3_117.

37. Karl Marx, *Capital*, vol. 1 (New York: Vintage, 1976); Brett Clark and John Bellamy Foster, "Marx's Ecology in the 21st Century," *World Review of Political Economy* 1, no. 1 (2010): 144.

38. Clark and Foster, "Marx's Ecology."

39. Solomon, "Kuils River Multiple."

40. Rupa and Patel, *Inflamed*, 350.

Future Tense

The Role of Race, Risk, and Environmental Justice

Helen Bond

The Smithsonian Science for Global Goals "Environmental Justice!" community research guide was developed by the Smithsonian Science Education Center and partners to heighten awareness of environmental justice that leads to informed action. The guide was implemented in two classrooms at Howard University Middle School of Mathematics and Science located in Washington, D.C., the capital city in the United States. The school serves an urban population of students between the ages of twelve and fourteen and is located on the campus of Howard University, a historically Black college and university (HBCU) located in the U.S. capital city. While HBCUs educate students from around the world, their principal mission is to educate African Americans. This study was conducted within two classrooms at Howard University Middle School of Mathematics and Science. The Smithsonian Science for Global Goals "Environmental Justice!" community research guide specifically centers justice within disadvantaged communities and includes activities linking personal and community identities, exploring relationality with the environment, and connecting environmental situations with health outcomes. Due to preexisting disparities, communities of color in the United States often face disproportionate environmental impacts. Children and youth are particularly vulnerable to the cumulative burden of environmental hazards.

This essay draws from pretest and posttest surveys, student interviews and artifacts to interrogate the role of race, risk, and environmental injustice. The research problematizes dominant narratives of environmental education as less inclusive and overly focused on exploitative forms of economic development. We hypothesize that with increased knowledge and understanding of environmental issues in their own community, African American students are more likely to perceive and understand risk and resilience in crisis contexts. This study is framed in crisis literature and contextualized in a world that has experienced a global pandemic amid human-induced climate change. Ignoring the impact of these phenomena would constitute an incomplete analysis. Any investigation of environmental injustice must be considered in tandem with the COVID-19 pandemic or as an important backdrop to it. Scholars have called for increased cross-disciplinary research that considers both the pandemic and the climate crisis as important context.

We examine the interconnections between risk, race, and environmental injustice by examining the nature of perceptions that urban middle school students have regarding the environment. Responding to risk is dependent on how we perceive and understand that risk. Risk communication and mitigation education during crises like COVID-19 and climate change will require accurate information, trust, and behavior change. How we perceive and understand risk can partly impact our willingness to change our behavior in response to it. Environmental denialism is often a response to the instability brought on by a lack of trust or knowledge, or the inability to perceive and calculate risk. However, we found no evidence of climate change denial in this study.

We ask students what motivates them to act or to engage in pro-environmental behavior (PEB), especially in an era fraught with crisis. We use the concept of "risk society" to focus on community resilience and impact. Community resilience is the capacity of individuals and households in a community to absorb shock, endure, and recover from multiple waves of crises. High levels of exposure to community violence is a known risk factor for a number of negative outcomes throughout childhood into adulthood. Both COVID-19 and the ongoing climate crises create disproportionate exposure and impact in African American communities and neighborhoods.[1]

The Role of Race, Risk, and Education in Crisis

When disasters occur, recovery depends on individuals who are part of larger supportive communities to bounce back. The community's ability to withstand the effects of shock often depends on preexisting conditions, levels of preparedness, and the agency to take informed action.[2] In order to facilitate resilience and higher levels of preparedness, state and local governments can identify communities where additional resources and information might be deployed to mitigate the impact. However, preparedness plans are only as effective as communities are to receive them. Communities that have been discriminated against, devastated by toxic pollution, denied resources, and redlined are less receptive to policy tools and have less capacity and trust needed to implement them.

For example, a 1987 report by the United Church of Christ Commission for Racial Justice found that toxic waste dumps were more likely to be located in minority neighborhoods in the United States.[3] Race was identified as a major factor in the decision-making process. The Commission for Racial Justice concluded that the hazardous waste exposed unknowing communities of African Americans, Hispanic Americans, Asian Americans, Pacific Islanders, and Native Americans to toxic waste. The Commission for Racial Justice partnered with poor and Black communities in Warren County, North Carolina, to protest locating a PCB disposal landfill near their communities. Similar racialized health inequities were also evidenced during the COVID-19 pandemic.

The Commission for Racial Justice 1987 findings are also supported by more recent data that shed light on patterns of systemic inequality. The United Nations Sustainable Development Solutions Network (UN SDSN) released "In the Red: The U.S. Failure to Deliver on a Promise of Racial Equality," an index that measures the achievement of the Sustainable Development Goals (SDGs) based on how well U.S. states deliver the goals to the least served racial groups. The findings show that on average, SDG delivery is highly unequal with white communities receiving resources and services at a rate approximately three times that of nonwhite communities.[4]

Likewise, the Commission for Racial Justice found that the voices of poor and minority communities were not included in the decision-making regarding the location of toxic waste sites in their communities. The environmental and climate movement has historically prioritized the needs and concerns of the white middle and upper classes in the United States. In "The Whiteness of Green: Racialization and Environmental Education," Sheelah McLean argues that whiteness continues to be the dominant narrative within environmental education in the Canadian context, much like the U.S. context. She further argues that environmental education curriculum often does not attempt to be antiracist, but instead adopts a postracial frame, where sustainable development is the solution for inequality and underdevelopment. Such a narrative is often centered around progress that legitimizes conquest and delegitimizes solutions that challenge racialized structures and systems.[5]

This postracial frame within certain sectors of the environmental movement community has persisted, albeit with a much stronger justice core. Its forerunner, the conservation movement, shifted from forest protection to an ecology frame of environmentalism and sustainable development. A question about forest protection was included in a survey to students as one way to tap into the lasting impact of the conservation movement on school curricula. A postracial or color-blind framework to the environmental movement can give the impression that certain racial and ethnic communities are less concerned about the environment and instead more concerned with racial discrimination. This singular reasoning denies the impact of multiple intersecting identities and overlapping crises.

Intersectionality is key to merging environmental justice with an antiracist framework. An intersectional frame alleviates tension between nature and racial justice by recognizing that both are integral to achieving sustainable outcomes. A merged frame combines solutions that are more structural (restricting waste sites in minority communities) with more individualistic actions (recycling and planting trees).[6] Merged framing recognizes that environmental justice is racial justice and racial justice is climate justice. Otherwise, racialized stereotypes may be used to justify environmental injustices that feed the narrative that minorities are less connected to nature, and therefore less deserving of claims of environmental degradation in their communities.

Similar nature-based racialized narratives emerged during the COVID-19 crisis. One was that that African Americans were genetically immune to the virus. Stories about mythologized immunity proliferated among social media, even as African Americans were beginning to contract and die of the virus at increasingly disproportionate numbers.[7] The numbers of deaths in Africa were initially used as evidence. Historian Rana Hogarth argues in *Medicalizing*

Blackness: Making Racial Difference in the Atlantic World, 1780–1840 that these myths of presupposed Black immunity were essentially justification claims designed to highlight racial difference and justify fitness for slavery.[8] Presupposed claims of immunity during COVID-19 justified Black people's fitness as essential workers shielded from the virus by the mere color of their skin. Race was deployed to communicate fitness for lower-skilled labor deemed essential during a time of crisis.

The concept of essential workers was also deployed during the COVID-19 pandemic. These workers were less likely to have a college degree, have less access to quality health care, and more likely to be Black, Hispanic, or other people of color. Despite their designation as essential, these low-income workers were concentrated among the poor, disenfranchised, and women and men of color.[9] These workers were required to assume high levels of risk. These risks included assaults to their health and sometimes person, financial risk, and familial risks by repeated exposure to a deadly virus. According to the American Psychological Association, about three in ten essential workers indicated that their mental health had suffered during the pandemic. More than half noted that they engaged in unhealthy eating and lifestyle habits during this time.[10] Essential workers were inducted into the Department of Labor's Hall of Honor in 2022 in recognition of their sacrifices during the pandemic.

We set out to discover how African American middle school students understand risk in relation to nature and to their communities. Misconceptions and inaccurate communication can emerge during a crisis threatening societal cohesion and heightening risk. This raises the importance of education that centers race and justice, not only for sustainable development, but also for disaster risk reduction education. In our study, we were particularly interested in how connected students felt to nature and nature-based solutions and what builds those connections.

We also wanted to measure how students might self-report or self-describe their relationship to the natural world using a variety of crisis scenarios, that is, scenarios that juxtapose progress against nature. Does building new roads justify cutting down trees? Our research demonstrated that these communities are concerned about environmental justice in their own and other communities. They are also concerned about health and economic injustices and injustices perpetrated by the U.S. criminal justice system that put them and their communities at further risk.

We finally argue that an understanding of crisis and risk management is particularly relevant here. Understanding crisis is needed to achieve the goal of societal and community resilience and cohesion. Community resilience is based on trust. This raises the question of what motivates action among youth and vulnerable minorities. How can trust be increased to leverage the strength of communities and youth for informed action? This happened sporadically during the COVID-19 pandemic. How might this happen for the ongoing crisis of climate change? What is the role of education in this process?

A crisis is not a point in time, but an emerging threat assessment that can further destabilize sustainable development. Crises are embedded in the conditions that give rise to them. Understanding the interrelated nature of these conditions (economic, social, environmental, and health) are central to this essay. The concept of crisis is based upon a risk society, that is, a society that is layered with emergent risks in areas such as the financial, political, technological, environmental, health, and global sectors.[11] All societies are risk societies, but may or may not transition into crisis. Some social scientists refer to these emergent risks as social and environmental disruptions. These disruptions can be short-term with long-range consequences like toxic smoke pollution from drifting wildfires. Long-term disruptions like the COVID-19 pandemic serve as ruptures to the fabric of society that have generational consequences. The concept of rupture is being used for specific and intense episodes of change.[12]

In an emergency, education functions as a humanitarian response. Even in a crisis such as the COVID-19 pandemic, education often remained a key priority for communities.[13] The expectation of participating in schooling is a routine that helps anchor children and families in their newfound circumstances. The consistency and predictability of going to school whether online or in person serves multiple purposes; the continuation of learning is important even if the format is different. Second, the curriculum can better serve students and their families if teachers and school personnel are prepared to help them mitigate risk and trauma. Disaster risk reduction education should be a part of the educational response so children and youth can better understand what is happening around them and stay safe.

The pandemic revealed the interconnectedness between health and well-being and the environment and racial justice. This essay describes how the COVID-19 pandemic was a disruptive event in a warming world. Using an intersectional lens to understanding the nature of risk, both crises highlight the need for disaster risk education and education for sustainable development. The danger of misinformation and lack of trust, preparedness, and critical thinking can heighten vulnerability and disaster risk. Disaster risk is a state of heightened vulnerability where the potential for loss of life, livelihoods, and access to basic needs, like water, food, education, and safety, is at its highest.[14]

Acquiring knowledge and skills is considered one of the most effective ways to reduce personal risk for further harm. Disaster risk education coupled with public health awareness, sex education, and risk communication can significantly impact people's readiness and response to emergencies. Resilient development is a long-term goal of disaster risk education. Resilient development provides children and families with preparedness skills to better anticipate and manage crises, but also to help recover from them. Vulnerable populations such as the elderly, people with disabilities, and children are often dependent on others for safety and security and can be especially at risk during an emergency. Disaster risk education is considered one of the most effective and innovative tools to increase the perception and understanding of risk among children and youth.[15] Maintaining the regularity of schooling during a disaster and after a disaster increases the relevance of education and its short- and long-term effectiveness.[16]

Research Methodology

Purpose and Participants

The Smithsonian Science Education Center (SSEC) in collaboration with the Howard University Middle School of Mathematics and Science conducted a study on the impact of the Smithsonian Science for Global Goals "Environmental Justice!" curriculum experience with approximately forty Black, Indigenous, People of Color (BIPOC) students. The school is a public charter school and committed to academic excellence, focusing on mathematics and science for students aged twelve to fourteen. The demographics of the school are approximately 98 percent African American. Students live in Washington, D.C., otherwise known as the District of Columbia—the formal designation for the U.S. capital city. Howard University Middle School of Mathematics and Science is located on the campus of Howard University in Washington, D.C. The area is also home to key branches of the U.S. federal government such as the White House, Supreme Court, and the Capitol. The Smithsonian is located nearby and is considered one of the largest museum and research institutions in the world.

The Smithsonian is a global institution focusing on science, conservation, culture, education, and the arts. The SSEC is part of the larger Smithsonian Institution that focuses on primary and secondary school education and how innovation, inclusion, and sustainability can increase access and help eliminate barriers. The SSEC collaborates with higher education institutions and communities in the United States and across the world, imagining new ways of solving problems to help make the planet more safe, just, and sustainable.[17]

Research Questions

The research questions focus on the impact of the "Environmental Justice!" curriculum on students in two classrooms at Howard University Middle School of Science and Mathematics. Their interest in science, technology, engineering, and mathematics (STEM)–related careers was also included in the study but is not explored here. A subsequent second set of questions that dealt with the impact of the BIPOC role models of the "Environmental Justice!" curriculum was also investigated in the original study. The part of the research study reported on here measures the degree of importance students place on certain environmental problems due to their perceived consequences and impacts. This is also a risk perception measurement.

The research study was conducted to answer the following interrelated research questions: How does the Smithsonian Science for Global Goals "Environmental Justice!" project enable BIPOC school students to:

- acquire knowledge and understanding of environmental issues in their own communities, as well as their interconnectedness to local, national, and global issues (cognitive);
- develop motivation and willingness to take informed actions (socioemotional);
- act effectively and responsibly, recognizing how our choices in one part of the world spill over, affecting people and the planet in other parts of the world (behavioral).

Instrument

The Environmental Concern Scale was used to measure the impacts and consequences that promote interest and problem solving around environmental phenomenon. Both variables—impact and consequences—are thought to prompt concern that may lead to increased agency and action.[18] The Environmental Concern Scale measured the degree of importance students place on certain environmental problems due to their perceived consequences and impacts. The scale ranges from 1 (not important) to 7 (extremely important).

Research Design and Analysis

We used a mixed-methods research (MMR) approach to study the impact of the Smithsonian Science for Global Goals "Environmental Justice!" community research guide. The guide is a student-facing resource designed to support youth in exploring environmental justice topics and problems within their local community. The guide includes eight hands-on science activities that help students learn how some environments can cause harm to both people, plants, and wildlife. Students identify a specific environmental issue and research its causes. To study this multidimensional phenomenon, we used the MMR approach that employed both qualitative and quantitative methods. Explanatory designs, that is, designs that are meant to explain how and to what degree something happened or changed or was impacted, often use qualitative methods after an initial quantitative engagement of the research (i.e., QUANT→QUAL).[19] MMR can draw from the strengths of both research traditions to help triangulate data and substantiate conclusions.[20]

For example, pretests were given and analyzed to help guide the follow-up focus group discussions. The focus group discussions were used to inform the semistructured interviews. Research activities included students engaging in lessons with the teacher and completing both pretest and posttest evaluations of their experience, including their knowledge, attitudes, and motivations concerning the environment, climate change, sustainability, and STEM. Activities also include completing pretest and posttest surveys, participating in a focus group session, and if selected, participating in a structured interview. Institutional review board approval and consent from both students and parents were obtained to participate in this project.

Since the research involved students and took place inside a school, convenience sampling was used to select participants. Convenience sampling is a type of purposive sampling where the researcher selects participants because they are readily available and willing to participate in the study.[21] Random samples—the holy grail of experimental designs—are generally not available in structured school settings. A pretest and posttest design were used to assess participants' knowledge, skills, attitudes, or perceptions relative to the prompt. An increase in knowledge, positive attitudes, or positive behavior change relative to the prompt is seen as evidence for a positive effect or relationship.[22] We compared pretest data to posttest data to determine the effect of an intervention. Demonstrating directionality in a quasi-experimental mixed-methods design indicates that the dependent variable (knowledge, skills, or attitude) was assessed before and after the intervention with an independent variable, such as the Smithsonian Science for Global Goals "Environmental Justice!" guide. Generalizability is limited to making associations between the intervention ("Environmental Justice!" curriculum) and outcomes. One cannot infer causality in this research study, but can infer association or correlation between the intervention (guide) and the outcomes (pre- and posttest scores).

Students participated in lessons taught by teachers, participated in select focus group sessions, and completed a pretest and posttest series. Students also had the option of participating in a semistructured interview. The pretest and posttest are used in this study for the purpose of measuring change resulting from a set of interventions. The non-self-contained focus groups served as a preliminary method to the structured interview to generate insights into what comes next. A conventional open coding method was used to code the unstructured data obtained from the interview and focus group forums.

The interviews were audio-recorded and transcribed. The interviews are only reported on to the extent that researchers observed for data saturation and informational redundancy, as well as for emerging codes and themes. Transcripts were read and interviews coded to help identify emerging themes. Classroom artifacts were also used to identify emerging themes or redundancies in the data. From these basic emerging themes, core themes were identified. Core themes are located through open or initial coding of the interviews. Once core themes were uncovered, the data was systematically coded for these key organizing constructs. Coding is defined as the systematic and rigorous process of sorting through data to locate and label codes to certain reoccurring words.[23] The researchers used three types of analysis: Qualtrics for survey development and descriptive analysis, and manual coding with pencil and paper.

Findings

This section reports on findings that measured the degree of importance students place on certain environmental phenomena due to their perceived consequences and impacts. This is one facet of environmental concern and connectedness and is considered to be a driver of PEB. We also measured students' knowledge and understanding of environmental issues and their willingness or agency to engage in PEB, that is, the willingness to take positive action on an environmental issue of perceived importance.[24] Three components make up PEB: the cognitive, behavioral, and socioemotional.[25] The cognitive component recognizes areas of concern and environmental threats. The behavioral component is the willingness to act on these issues and concerns.[26] The socioemotional factor is reflected in dispositions that favor the environment.[27] Kaiser Florian, Sybille Wölfing, and Urs Fuhrer discuss a theory of reasoned action that further defines pro-environmental behavior. These six categories include factual knowledge, attitude toward behavior, social and moral values, subjective norms, behavior intention, and behavior.[28] These parameters informed our understanding of PEB, but did not limit it. We intentionally focused on a broad, more interdisciplinary definition that is more adaptable to risk and crisis contexts.[29]

Environmental Connectedness and Concern

Environmental connectedness is one's relationship to the natural environment and is also an important driver of PEB.[30] Environmental connectedness, that is, how connected we feel to nature and/or our community, can be predictive of engaging in PEB.[31] Feeling connected to nature is related to self-transcendence, that is, valuing the welfare of others over your own. Self-transcendence is highly correlated with PEB. Narcissism, bias, and self-interest also predict less pro-environmental action or PEB.[32]

The Environmental Concern Scale is designed to capture these relationships that underlie PEB. The scale measures what drives people to action. People are generally concerned about environmental problems because of the negative consequences that they perceive can result from them. The Environmental Concern Scale attempts to measure the degree of importance students place on certain environmental problems due to their perceived consequences and impacts. The scale ranges from 1 star (little concern) to 7 stars (very concerned). The Environmental Concern Scale taps into what concerns people and why regarding the environment. It is worth noting that not all impacts and consequences are of equal concern. People differ in what they find concerning and what motivates them to take action. This was the case with the eighth-grade middle school students in this study.

Results

The Environmental Concern Scale table summarizes the results from the Environmental Concern Scale prompt asking students "I am concerned about environmental problems because of the consequences for (me, my future, people, the planet, my children, my health, my community, ocean and marine life, animals, forests, birds, plants, and my lifestyle)."

Eighth-grade students ages twelve to fourteen consistently indicated that they were very concerned about environmental problems because of the consequences for the planet, my health, my children, my future, and me by ranking each of these environmental values (EVs) with seven stars. These five EVs received the seven-star ranking by over 50 percent of the students in both the pre- and posttest (see the table).

Black Bodies

In both pretests and posttests, African American middle school students indicated with a high degree of confidence that they were most concerned about the environment because of the impact on them. This includes their lives, their bodies, including their health and well-being, and any future offspring. This high degree of concern also extended to the planet and the animals, plants, and people that populate it. Despite the misperception that Black communities have less interest in the environment, African Americans have deep roots fighting for

THE ENVIRONMENTAL CONCERN SCALE

I AM CONCERNED ABOUT ENVIRONMENTAL PROBLEMS BECAUSE OF THE CONSEQUENCES FOR:			
1 STAR (NOT CONCERNED) TO 7 STARS (VERY CONCERNED)			
ENVIRONMENTAL VALUES (EVS) RATED 7 STARS	*PRE	ENVIRONMENTAL VALUES (EVS) RATED 7 STARS	*POST
Planet	78%	Planet +	82%
My Children	74%	My Health +	81%
My Health	74%	Me +	76%
My Future	56%	My Future +	75%
Me	54%	My Children --	69%
Ocean and Marine Life	38%	Animals +	63%
Plants	37%	Forests +	56%
My Lifestyle	37%	Plants +	56%
Forests	37%	My Lifestyle +	50%
Animals	34%	Birds +	50%
Birds	26%	Ocean and Marine Life +	50%
My Community	15%	My Community +	44%
People	15%	People +	41%

* Percentages are rounded
+ Indicates an increase in value from pretest to posttest
-- Indicates a decrease in value from pretest to posttest
Population: N27

environmental justice. Fannie Lou Hamer is an example of how the preservation of the Black body was deeply entrenched in the fight for voting rights, gender equality, racial justice, and environmental justice.[33]

Hamer was one of twenty children born on the Marlow plantation in the Mississippi Delta. She established the Freedom Farms Cooperative in rural Sunflower County, Mississippi, in 1969. Hamer's goals were to create safe spaces to grow food, while advancing the goals of the larger movement.[34] In Priscilla McCutcheon's "Fannie Lou Hamer's Freedom Farms and Black Agrarian Geographies," she argues that Hamer's Freedom Farm was a Black radical space operating at three levels: saving the black body, the geography of the farm, and the southern agrarian landscape.[35]

A national study addressing pro-environmental inclinations of Hispanic Americans and African Americans found that after controlling for racial and demographic differences, both communities were found to be as concerned with environmental justice in comparison with the majority group.[36] In this study 96 percent of students in both pre- and posttests agreed or strongly agreed that humans are severely abusing the environment. The belief in human-induced climate change is a strong predictor of environmental awareness and PEB. Figure 1 includes pretest data results, which correlate with posttest data.

Environmental injustice in some sectors has been narrowly conceived as violations to the planet but not the people. The focus is often on illegal waste management in poor and minority communities. The 1987 report by the United Church of Christ Commission for Racial Justice helps reframe environmental justice as racial justice. Willie Jamaal Wright argues in "As Above, So Below: Anti-Black Violence as Environmental Racism" that the destruction and gentrification of environmental habitats is a feed for anti-Blackness. He appeals to the reader that environmental racism is a mutual devaluation and destruction of Black bodies.[37]

After engaging with the "Environmental Justice!" guide and with female BIPOC environmental justice leaders as role models, students further expanded their circle of concern to include oceans, forests, birds, and plants and animals. These EVs were rated higher (given more stars) after the intervention ("Environmental Justice!" curriculum). Students' environmental concerns were analogous to Maslow's hierarchy of needs. Shaped like a pyramid, Maslow's hierarchy has six drivers of human motivation. Physiological needs such as food and water form the base. The next domain is safety and belonging, and the peak of the pyramid is self-actualization. Maslow's theory portends that basic primordial needs like food and water must be met before fully realizing more conceptual needs like self-esteem and self-actualization.[38]

Calculating Risk

Risk perception plays a role here. Students are calculating risk when they base their concern about environmental problems on the consequences for their

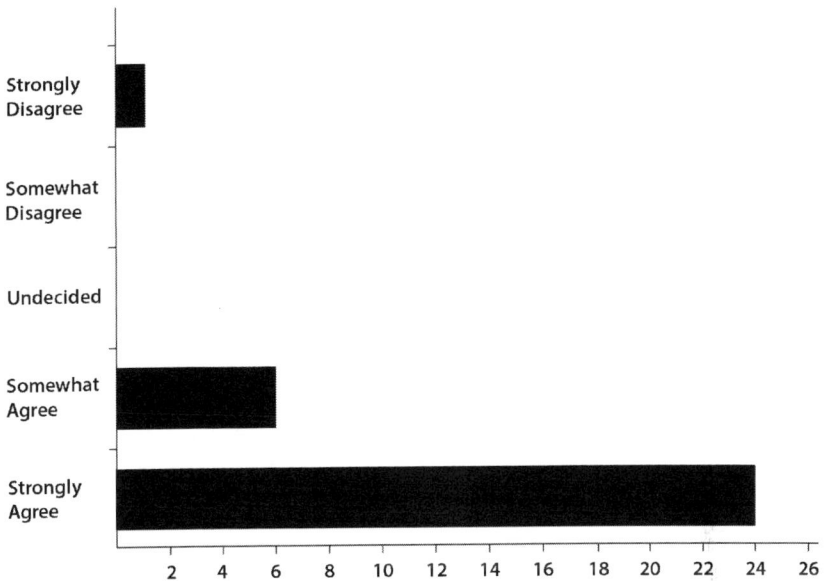

FIGURE 1. Humans are severely abusing the environment. * Pretest data Population: N31.

personhood, their health, and their futures. These are assessments by young African American students who are calculating risk based on their safety and well-being. They include concern for the planet if a future for them is even possible. They recognized that their future was inextricably linked to planet Earth. When asked, "If things don't change, we will have a big disaster in the environment soon," on a Likert scale survey, approximately 78 percent of students strongly agreed. This increased to 85 percent on the posttest assessment—a full 7 percent increase. Students in this study are aware of police violence in their communities and the tragedy of death from COVID-19, which hit Black and poor communities especially hard. In 2021, African Americans in the United States were nearly fourteen times more likely to die of gun violence than their white counterparts.[39]

Risk perception is partly based on proximity, vulnerability, and visions for the future. The closer you are to the threat, real or perceived, the more likely you are to act. Closeness or vulnerability can refer to geographical distance, but can also refer to a social and cultural distance. Essential workers often held more public-facing employment, which placed them in higher proximity to the virus. While other factors weigh in, like trust, self-efficacy, social norms, and accurate information, people often base their risk calculations on how it impacts them and those closest to them.[40]

This aligns with the protection motivation theory. This health-protective behavior theory suggests that people act under these three conditions: when a threat is perceived to be high, when they have sufficient levels of agency and self-efficacy to respond, and when the costs for doing so are low.[41] This was confirmed in a study on risk perceptions and health behaviors during COVID-19 in the U.S. context. Researchers found support for the claim that many people changed their behavior in light of the ongoing threat level and public health guidelines.[42]

As the pandemic became more politicized, risk calculations changed in response. African Americans' support for public health measures such as mask and glove wearing generally remained strong. However, vaccine uptake was still low among young Black males due to social norms and lack of trust in the medical community. Vaccine hesitancy was driven by medical mistrust in the health-care establishment due to egregious acts of medical malpractice against African Americans, such as the infamous stories of Henrietta Lacks and the Tuskegee experiments. All were rooted in structural discrimination that continues to exploit health inequities in many Black communities. To address vaccine hesitancy among young Black males, a preexisting Tough Talks application was modified to include trusted messengers, engaging scenarios that promoted good decision-making skills, and risk awareness training.[43]

Xin Li, Zhenhui Liu, and Tena Wuyun studied the relationship between students' EVs and PEB using a values-belief-norm framework. The results indicate that EVs can be predictive of PEB among young adults and that risk perception, moral anger, and the lack of trust can play a decisive role.[44] Risk calculation and moral equivalencies are thought to be predictive of agency that leads to action or engaging in environmentally friendly behavior or PEB.[45] PEB is defined as a form of prosocial behavior prompted by internal values informed by the assessment of risk.[46] Narcissism and egoistic values are negatively related to environmental concern. PEB is also a form of risk assessment. The more risk one perceives, the more likely a behavior change will occur.

Future Tense

Throughout the study students expressed concern for the environment because of the consequences of its destruction on their lives and futures. This concern extended beyond their lives to those of their offspring. In both pre- and posttests, students indicated that they were concerned about environmental problems because of the consequences for them, their health, their children, and the planet (see the table). Rather than narcissism, students are expressing aspects of future orientation. Future orientation refers to being hopeful. It also refers to being motivated by and planning for a brighter future. Another dimension of future orientation is the belief that one can set goals and have a realistic expectation that they can be achieved.[47]

Adolescence is a period where the ability to imagine a future is made possible by cognitive changes in the brain and physical changes to the body. Being future oriented has been identified as a protective factor for youth in both low- and high-risk environments.[48] The ability to overcome adverse circumstances requires a recognition of the past and an outlook toward the future.[49] Adolescents that can imagine a more positive future may perceive and respond to risk differently. Researchers find that they tend to have fewer externalizing behaviors, which may decrease their likelihood of engaging in high-risk activities. Future orientation is also thought to facilitate the transition to adulthood. Fatalism, which is the opposite of hope, has been linked to negative mental health outcomes.[50]

Racism and discrimination in the lives of BIPOC youth undercut their ability to be future positive. Black adolescents are exposed to racial discrimination in the United States at levels that can thwart their ability to control and plan for a successful future. According to a National Survey of Children's Health, BIPOC children and youth experienced discrimination in the United States at high rates that increased from 2016 to 2020.[51] The accumulation of risk and adversities stymies the ability to both plan on and execute one's future. While more research needs to be done on how Black youth perceive their future, this study provides some evidence that they are indeed thinking about their future and the future of their offspring.

For the purposes of this study, we characterize the future orientation of the students we studied as "future tense." Future tense is characterized by futures thinking, but with questionable expectations that the future envisioned is possible or even desirable—in the current circumstances. This renders future work for these students less usable. Future tense adolescents have two envisioning tasks. One is to stake out a future that is possible under the current restraints (racism, discrimination, homophobia, poverty). The second task is to envision a world without these restraints; a world where your desirable future is indeed achievable given the necessary inputs.[52] While the Black adolescents that participated in this study are projecting, planning, and thinking about the future, their aspirations are nested in overlapping systems of racial discrimination and hierarchy.[53] Despite these constraints, the data in this study suggest that African American adolescents still have a sense of purpose for their future selves and their progeny.

Conclusion

The Smithsonian Science for Global Goals "Environmental Justice!" community research guide helped students develop the knowledge, skills, and dispositions needed to take informed action on environmental issues within their communities. Heidi Gibson, in "From Ideas to Action: Transforming Learning to Inspire Action on Critical Global Issues," emphasizes how students can bring their own unique community knowledge and perspectives to the learning experience to help guide action-taking.[54]

Critical sustainability mindsets also include good risk perception, open-mindedness, a respect for diversity, and being future oriented. Future tense is a mitigation strategy for envisioning futures not yet possible. It doubles the work of creating usable futures for BIPOC youth. Our research demonstrates that implementing environmental education in the classroom should center antiracism, justice, and sustainability, as well as future thinking. Situating this learning in the context of the SDGs and child-centered disaster risk reduction empowers students to make connections between the global and local and to mitigate risk. It helps them to recognize how the choices we make impact not only us, but also people in faraway places.

Acquiring these critical mindsets is also considered one of the most effective ways to understand and mitigate risk.[55] Reducing risk in communities trying to manage multiple ongoing crises, such as racism, climate change, epidemics, and disasters, is critical to advancing sustainable development that is truly sustainable and just. It is also important to adopt disaster risk reduction education by equipping students with the knowledge and skills to calculate risks, while taking informed action. As climate change continues to accelerate, mitigating and managing risk will become increasingly important.[56]

NOTES

1. Briana Woods-Jaeger et al., "Building a Contextually-Relevant Understanding of Resilience among African American Youth Exposed to Community Violence," *Behavioral Medicine* 46, no. 3/4 (July 2020): 335, doi: 10.1080/08964289.2020.1725865.

2. Rajib Shaw, Aiko Sakurai, and Yukihiko Oikawa, "New Realization of Disaster Risk Reduction Education in the Context of a Global Pandemic: Lessons from Japan," *International Journal of Disaster Risk Science* 12, no. 4 (August 2021): 578–80, doi: 10.1007/s13753-021-00337-7.

3. "Toxic Waste and Race in the United States: A National Report on the Racial and Socio-Economic Characteristics of Communities with Hazardous Waste Sites, 1982," *UNC Libraries*, https://exhibits.lib.unc.edu/items/show/7441.

4. Alaina Lynch, Helen Bond, and Jeffrey Sachs, "In the Red: The U.S. Failure to Deliver on a Promise of Racial Equality," Sustainable Development Solutions Network, May 5, 2021, https://www.sdsnusa.org/publication/inthered.

5. Sheelah McLean, "The Whiteness of Green: Racialization and Environmental Education," *Canadian Geographer* 57, no. 3 (Autumn 2013): 354, doi: 10.1111/cag.12025.

6. Milanika S. Turner, "Climate Change Hazards + Social Vulnerability = A Recipe for Disaster," *Generations: Journal of the American Society on Aging* 46, no. 2 (July 1, 2022): 4.

7. Cato T. Laurencin and Aneesah McClinton, "The COVID-19 Pandemic: A Call to Action to Identify and Address Racial and Ethnic Disparities," *Journal of Racial and Ethnic Health Disparities* 7, no. 3 (2020): 399, https://doi.org/10.1007/s40615-020-00756-0.

8. Rana Hogarth, *Medicalizing Blackness: Making Racial Difference in the Atlantic World 1780–1840* (Chapel Hill: University of North Carolina Press, 2017), 29.

9. "No Worker Left Behind: Supporting Essential Workers," Hearing before the Committee on Oversight and Reform, 116th Cong., 2d Sess., June 10, 2020, 1–3, https://www.govinfo.gov/content/pkg/CHRG-116hhrg40842/html/CHRG-116hhrg40842.htm.

10. "Essential Workers More Likely to be Diagnosed with a Mental Health Disorder during Pandemic," American Psychological Association, March 21, 2021, https://www.apa.org/news/press/releases/stress/2021/one-year-pandemic-stress-essential#:~:text=Essential%20workers%20were%20more%20than,9%25).

11. V. G. Voronkova, "Risk Society as a Result Crisis of Modern Civilization in the Global Dimension," *Humanities Bulletin of Zaporizhzhe State Engineering Academy*, no. 58 (November 2014): 14.

12. Sango Mahanty et al., "Rupture: Towards a Critical, Emplaced, and Experiential View of Nature-Society Crisis," *Dialogues in Human Geography* 13, no. 2 (2023): 177, doi:10.1177/20438206221138057.

13. Ian Phil Canlas, "Three Decades of Disaster Risk Reduction Education: A Bibliometric Study," *Natural Hazards Research* 3, no. 2 (June 2023): 329, doi:10.1016/j.nhres.2023.02.007.

14. Geetanjali Saini, Monica H. Swahn, and Ritu Aneja, "Disentangling the Coronavirus: Disease 2019 Health Disparities in African Americans: Biological, Environmental, and Social Factors," *Open Forum Infection U.S. Diseases* 8, no. 3 (March 2021): 7, doi:10.1093/ofid/ofab064.

15. Rajib Shaw, Koichi Shiwaku, and Yukiko Takeuchi, *Disaster Education: Community, Environment and Disaster Risk Management* (Bingley: Emerald Group, 2011), 32.

16. Weiwei Zhang and Deepthi Kolady, "The COVID-19 Pandemic's Unequal Socioeconomic Impacts on Minority Groups in the United States," *Demographic Research* 47 (2022): 1019.

17. "Smithsonian Science for Global Goals Project," Smithsonian Science Education Center, July 15, 2023, https://ssec.si.edu/global-goals.

18. Christie P. Karpiak and Galen L. Baril, "Moral Reasoning and Concern for the Environment," *Journal of Environmental Psychology* 28, no. 3 (2008): 207, doi:10.1016/j.jenvp.2007.12.001.

19. Annette Kluge et al., "Investigating Unlearning and Forgetting in Organizations: Research Methods, Designs and Implications," *The Learning Organization* 26, no. 5 (2019): 518, doi:10.1108/TLO-09-2018-0146.

20. David E. Houchins, John H. Hitchcock, and Maureen A. Conroy, "A Framework for Approaching Mixed Methods Intervention Research to Address the Emotional and Behavioral Health Needs of Children," *Behavioral Disorders* 48, no. 3 (May 2023): 161, doi:10.1177/01987429221131279.

21. Samuel J. Stratton, "Quasi-Experimental Design (Pre-Test and Post-Test Studies) in Prehospital and Disaster Research," *Prehospital and Disaster Medicine* 34, no. 6 (2019): 573, doi:10.1017/S1049023X19005053.

22. Kristine Bakkemo Kostøl and Kari Beate Remmen, "A Qualitative Study of Teachers' and Students' Experiences with a Context-Based Curriculum Unit Designed in Collaboration with STEM Professionals and Science Educators," *Disciplinary and Interdisciplinary Science Education Research* 4, no. 1 (2022): 12, doi:10.1186/s43031-022-00066-x.

23. Johnny Saldana, *Fundamentals of Qualitative Research: Understanding Qualitative Research* (New York: Oxford University Press, 2011), 66–70.
24. Paul C. Stern, "Toward a Coherent Theory of Environmentally Significant Behavior," *Journal of Social Issues* 56, no. 3 (September 2000): 408, doi:10.1111/0022-4537.00175.
25. Olivia Koon, Ricky Y. K. Chan, and Piyush Sharma, "Moderating Effects of Socio-Cultural Values on Pro-Environmental Behaviors," *Marketing Intelligence & Planning* 38, no. 5 (2020): 605, doi:10.1108/MIP-10-2019-0534.
26. Eero Olli, Gunnar Grendstad, and Dag Wollebaek, "Correlates of Environmental Behaviors: Bringing Back a Social Context," *Environment and Behavior* 33, no. 2 (2001): 185, https://doi.org/10.1177/0013916501332002.
27. Daphne Goldman et al., "Influence of 'Green School Certification' on Students' Environmental Literacy and Adoption of Sustainable Practice by Schools," *Journal of Cleaner Production* 183 (2018): 1300, doi:10.1016/j.jclepro.2018.02.176.
28. Florian G. Kaiser, Sybille Wölfing, and Urs Fuhrer, "Environmental Attitude and Ecological Behaviour," *Journal of Environmental Psychology* 19, no. 1 (1999): 3, doi.org/10.1006/jevp.1998.0107.
29. Ediyanto Ediyanto et al., "Sustainable Instrument Development in Educational Research," *Discourse & Communication for Sustainable Education* 13, no. 1 (June 2022): 37, doi:10.2478/dcse-2022-0004.
30. Joanne Dono, Janine Webb, and Ben Richardson, "The Relationship between Environmental Activism, Pro-Environmental Behaviour and Social Identity," *Journal of Environmental Psychology* 30, no. 2 (2010): 185, doi:10.1016/j.jenvp.2009.11.006.
31. Wesley P. Schultz and Lynette C. Zelezny, "Values as Predictors of Environmental Attitudes: Evidence for Consistency across 14 Countries," *Journal of Environmental Psychology* 19, no. 3 (1999): 264, doi:10.1006/jevp.1999.0129.
32. Dono, Webb, and Richardson, "The Relationship between Environmental Activism, Pro-Environmental Behaviour and Social Identity," 178.
33. Nicholas Targ, "Human Rights Hero: Fannie Lou Hamer (1917–1977)," *Human Rights* 32, no. 2 (2005): 25.
34. Janice D. Hamlet, "Fannie Lou Hamer: The Unquenchable Spirit of the Civil Rights Movement," *Journal of Black Studies* 26, no. 5 (1996): 572.
35. Priscilla McCutcheon, "Fannie Lou Hamer's Freedom Farms and Black Agrarian Geographies," *Antipode* 51, no. 1 (January 2019): 208, doi:10.1111/anti.12500.
36. Troy Elias et al., "Understanding Climate Change Perceptions and Attitudes across Racial/Ethnic Groups," *Howard Journal of Communications* 30, no. 1 (January 2019): 39, doi:10.1080/10646175.2018.1439420.

37. Willie Jamaal Wright, "As Above, So Below: Anti-Black Violence as Environmental Racism," *Antipode* 53, no. 3 (May 2021): 791–92, doi:10.1111/anti.12425.

38. Katja Košir and Katarina Habe, "Analysis of Learning Environment Factors Based on Maslow's Hierarchy of Needs," *Revija Za Elementarno Izobraževanje* 6, no. 2–3 (2013): 178.

39. Ari Davis, Rose Kim, and Cassandra Crifasi, "A Year in Review: 2021 Gun Deaths in the U.S.," Johns Hopkins Center for Gun Violence Solutions, Johns Hopkins Bloomberg School of Public Health, June 6, 2023, https://publichealth.jhu.edu/2023/new-report-highlights-us-2021-gun-related-deaths-for-second-straight-year-us-firearm-fatalities-reached-record-highs.

40. Roxane Cohen Silver, E. Alison Holman, and Dana Rose Garfin, "Coping with Cascading Collective Traumas in the United States," *Nature Human Behavior* 5 (2021): 4, https://doi.org/10.1038/s41562-020-00981-x.

41. Steven Prentice-Dunn and Ronald W. Rogers, "Protection Motivation Theory and Preventive Health: Beyond the Health Belief Model," *Health Education Research* 1, no. 3 (1986): 160, https://doi.org/10.1093/her/1.3.153.

42. Dana Rose Garfin et al., "Risk Perceptions and Health Behaviors as COVID-19 Emerged in the United States: Results from a Probability-Based Nationally Representative Sample," in "Risk Perception, Decision Making, and Risk Communication in the Time of COVID-19," ed. Susan Joslyn, Gale M. Sinatra, and Daniel G. Morrow, special issue, *Journal of Experimental Psychology: Applied* 27, no. 4 (December 2021): 594, doi:10.1037/xap0000374.supp.

43. Henna Budhwani et al., "Tough Talks COVID-19 Digital Health Intervention for Vaccine Hesitancy among Black Young Adults: Protocol for a Hybrid Type 1 Effectiveness Implementation Randomized Controlled Trial," *JMIR Research Protocols* 12 (2023): 13, e41240, doi:10.2196/41240.

44. Xin Li, Zhenhui Liu, and Tena Wuyun, "Environmental Value and Pro-Environmental Behavior among Young Adults: The Mediating Role of Risk Perception and Moral Anger," *Frontiers in Psychology* 13 (2022): doi:10.3389/fpsyg.2022.771421.

45. Darko Hinić, "Should I Be Bothered or Not? Development of the Environmental Attitudes Scale (EAS)," *Primenjena Psihologija* 15, no. 3 (July 2022): 425, doi:10.19090/pp.v15i3.2385.

46. Gertrúd M. Keresztes and Ibolya Kotta, "From Perceiving the Risk of Climate Change to Pro-Environmental Behavior," *Acta Didactica Napocensia* 14, no. 2 (2021): 127.

47. Sarah Lindstrom Johnson, Robert W. Blum, and Tina L. Cheng, "Future Orientation: A Construct with Implications for Adolescent Health and Wellbeing," *International Journal*

of Adolescent Medicine and Health 26, no. 4 (2014): 463, doi: 10.1515/ijamh-2013-0333.

48. Silver, Holman, and Garfin, "Coping with Cascading Collective Traumas in the United States," 4–5.

49. Jun Sung Hong et al., "Future Orientation and Adverse Outcomes of Peer Victimization among African American Adolescents," *Journal of Aggression, Maltreatment & Trauma* 30, no. 4 (April 2021): 528, doi: 10.1080/10926771.2020.1759747.

50. Sarah Lindstrom Johnson, Elise Pas, and Catherine Bradshaw, "Understanding the Association between School Climate and Future Orientation," *Journal of Youth & Adolescence* 45, no. 8 (August 2016): 1585, doi: 10.1007/s10964-015-0321-1.

51. Peter Cabrera, Wendy Auslander, and Michael Polgar, "Future Orientation of Adolescents in Foster Care: Relationship to Trauma, Mental Health, and HIV Risk Behaviors," *Journal of Child & Adolescent Trauma* 2, no. 4 (October 2009): 272–73, doi: 10.1080/19361520903317311.

52. Gloria Marsay, "A Hope-Based Future Orientation Intervention to Arrest Adversity," *South African Journal of Education* 40, no. 1 (2020): 7.

53. Jacqueline O. Moses, Miguel T. Villodas, and Feion Villodas, "Black and Proud: The Role of Ethnic-Racial Identity in the Development of Future Expectations among At-Risk Adolescents," *Cultural Diversity and Ethnic Minority Psychology* 26, no. 1 (January 2020): 112–13, doi: 10.1037/cdp0000273.

54. Heidi Gibson, "From Ideas to Action: Transforming Learning to Inspire Action on Critical Global Issues," Smithsonian Institution Scholarly Press, Smithsonian Institution, August 18, 2021, https://scholarlypress.si.edu/store/life-sciences-biodiversity/ideas-action-transforming-learning-inspire-action/.

55. Viktor Gecas and Michael L. Schwalbe, "Beyond the Looking-Glass Self: Social Structure and Efficacy-Based Self-Esteem," *Social Psychology Quarterly* 46, no. 2 (1983): 78.

56. Troy Elias, "The Impact of Media Use, Identity, and Pro-Environmental Orientations on Racial/Ethnic Groups' Attitudes toward Ecobranding," *Howard Journal of Communications* 31, no. 1 (January 2020): 99, doi: 10.1080/10646175.2019.1649761.

Decolonizing Environmental Justice

Centering Black and Indigenous Solutions to the Climate Crisis

Autumn Asher BlackDeer and Sierra Roach Coye

Environmental injustices such as the climate crisis are inextricably tied to the ongoing processes of colonialism, dispossession, and racial capitalism. Whiteness has been produced through the violence of colonialism and continues to erase Black and Indigenous communities and our histories from the land. Results of colonial whiteness render conceptions of the wilderness as an empty space, entitling white people to occupy and claim the environment as a white space. Thus, so-called solutions to the white climate crisis are grounded in colonial thinking and perpetuate systemic harm against Black and Indigenous communities. Western colonial logic has long been the prevailing voice in the climate crisis discussion. Western capitalist society views the natural world as a commodity, a property that can be owned, or a resource to be exploited. Further, the climate crisis itself centers humans, again positioning this Western hierarchy rather than placing oneself in relation with the land. Overall, these Western colonial conceptions of environmental justice fail to encapsulate what Black and Indigenous communities consider justice. Typical formulations are not translatable to Black and Native communities without grounding in our own philosophies, ontologies, epistemologies, and worldviews.

The present work centers Indigenous cosmologies and abolition ecologies as a direct response to decolonize the Environmental Justice movement. For example, Indigenous ontologies in the water justice movement posit understanding water as a living entity that ensures the well-being of life, again not a mere commodity or property. Grounded in our lived experiences as queer Black and Native scholars in conversation with the scholarly literature, this work will (1) delineate historic impacts of colonialism on Black and Indigenous communities, (2) demonstrate the continued coloniality within the present-day Environmental Justice movement, and (3) introduce Indigenous cosmologies and abolition ecologies as solutions to decolonize environmental justice. Ultimately, there can be no environmental justice without Black liberation and Indigenous sovereignty.

Introduction

Over 80 percent of global biodiversity occurs within Indigenous territories, which we know are constantly under attack by state and corporate development.[1] In settler colonial states like the United States, there is ongoing marginalization of Indigenous people and disregard for Indigenous environmental knowledge.[2] White supremacist practices have deeply institutionalized control over land; it is imperative to name and dismantle these existing institutions that uphold these oppressive traditions.[3] Whiteness has been produced historically through violent land grabs from Indigenous communities, yet little attention has been paid to racialized property relations and the resulting ability of communities to thrive beyond mere survival.[4] Anthropocentric climate change has been identified as the defining issue of our time by the world's leading experts, with many now opting to use "climate crisis" over climate change to stress the urgency with which we must act to achieve a sustainable future, acknowledging the need to recenter Indigenous relations to land and address the horrors of settler colonialism and racism.[5]

Social injustices experienced by Black and Native communities, such as slavery, discrimination, genocide, and land theft, are also forms of environmental injustice.[6] Several have called for analyses that do not employ the Black body by ignoring Black people, further dehumanizing people in our scholarship in the pursuit of critiquing racialized violence.[7] This discourse is similar in studies of Indigenous communities, often reporting only on disparities and the amount of ancestors in the ground rather than our resilience and survivance. As queer critical decolonial and abolitionist scholars, we approach this work from our lived experience and academic expertise. The first author identifies as a queer Indigenous decolonial scholar-activist from the Southern Cheyenne Nation. The second author is a queer Black scholar and practitioner grounded in liberatory Afrocentric frameworks. This work has grown from our collective world-building collaborations and endeavors that place our fields of decolonization and abolition in conversation with one another.

Settler Colonialism

Beginning in 1492 with colonizer arrival, it is estimated that their colonial campaigns killed more than fifty-five million Indigenous peoples in what is known today as the United States.[8] Colonization began with the exploitation of Native lands and resources in order for settlers to gain wealth and power while subsequently making Natives dependent upon state resources.[9] European settlers and the growing federal government displaced and ultimately removed Native people from their land. Natives were forcibly relocated to reservations often far from their traditional tribal lands with the Indian Relocation Act of 1956.[10] This forced relocation to urban areas brought homelessness, unemployment, poverty, and a lack of a cultural base or community.[11] This act ultimately set the stage for the vast differences we see between reservation and urban Native populations today.

Once Natives were removed, the government then parceled land into allotments and sold tiny pieces of the reservation back to individual Natives in hopes of encouraging agriculture and ultimately assimilation. The remaining reservation lands were sold to settlers at a profit for the federal government. Such acts of greed and supporting entitlement of European settlers led to multiple attempts at overhauling tribal institutions, belief systems and practices, and traditional ways of life.[12] Natives have had little control over their own resources and economic situations since the removal period.

Settler colonialism is an ever-present living thing, not a historic relic.[13] Settler colonialism transformed from a multitude of relations across frontiers to now culminate into a hegemonic project of ethnic and ecological cleansing to make way for colonial settlement.[14] The process of colonization moved focus away from valuing relationships between humans and other beings and instead shifted toward resource extraction through exclusive individualized landownership; issues such as these demonstrate the lasting legacy of colonization, domination, and oppression.[15]

Present-day accounts of pre-European habitation romanticize Indigenous peoples and our relation to the land. These historicized narratives of settlement within larger, racist narratives of migration and cultural exchange claim that Native history is a component of settler colonial cultural lineage, rendering Indigenous people a small blip on the colonial agenda throughout time, disregarding the treaty process, and ultimately portraying treaties as exclusive ownership agreements rather than the joint use and occupation understandings they were meant to be.[16] Ultimately today we know that the United States has broken every single treaty ever signed with Native Nations.

Since the Revolutionary War, the United States has systematically limited Native economic and political self-determination through the ultimate pursuit of total eradication and removal.[17] Colonization and the intensification of settler agriculture quickly led to ecological issues, namely poor sanitation and disease. These changes in the ecosystem were not regulated by the new settler regime; water deteriorated, the local economy shifted to trade as the primary means of accessing goods, rather than the local hunting, fishing, and gathering practices that were previously the norm.[18]

The succession of Indigenous lands to colonizers caused a cascade of extinction events for animal, vegetative, and geological life, many of which depended on Indigenous stewardship and care; this ecological destruction is a form of genocide: ecocide.[19] The cyclical and interdependent relationship between Indigenous peoples and the land meant that the decline in one meant a decline was simultaneously triggered in the other.[20] This phenomenon is encapsulated within the two systematic tipping points—relational and ecological—that describe the concurrent decline of Native cultures and the decline of nonhuman life due to climate change; these climate-related dangers are inseparable from the absence of respect for relational qualities—the disruption of Indigenous cultural practices and kinship with the land is central to understanding climate change and related risks.[21]

Uneven racial development has produced much of the foundation of U.S. history, with too many of these movements remaining absent from our collective geography and political ecological imagination particularly when considering the interconnected ways that property relations are directly enmeshed in broader environmental politics, enforcing ongoing harm and suffering through racial capitalism.[22] Writing on whiteness as property allows settler colonial governments to justify the process of Indigenous land dispossession via an ideology of improvement, often citing the lack of cultivation.[23] Indigenous persistence and presence on the land continues to be contested by settler fantasies; the continued existence of white supremacist logics perpetuates uneven racial development in land and property relations as settler relationships with land are motivated by pursuits of wealth through modes of domination and exploitation.[24]

The formation of the United States is territorially based in forms of oppression and violence against Native Nations and communities of color.[25] The project of slavery-based settler colonialism is a settler society based on racism, whose material well-being is entirely dependent on the expropriation of bodies and land justified by white supremacy.[26] So many sources of insight look for connections between colonization and racist ideology and environmental contradictions.[27] Mythologies of innocence are an integral aspect of the social fabric of settler society as many environmental education programs and ecojustice scholars alike problematically invite students to reconnect with the land without analyzing the violent history of colonization and white settlers' illegitimate occupation of Indigenous lands.[28] The history of colonial violence must be silenced to maintain a national identity of innocence; distancing whiteness from colonialism allows white settlers to imagine they reside within a postracial society.[29] The present work disrupts that silence.

Coloniality in Climate Crisis

Utilizing the historical context of the Indigenous and Black American experience, the subjugation of the Indigenous and Black body in connection with the subjugation of land has been a long-standing tradition. The United States is a country born out of genocide, for both Indigenous and Black peoples. The topic of environmental justice is always discussed as allowing all people regardless of race, national origin, or income to be involved in the development, implementation, and enforcement of laws, regulations, and policies.[30] However, if we were to enact change within environmental spaces, we would be highlighting Indigenous sovereignty and Black liberation to inform ways of engaging with the natural world. The entire definition outlined by the U.S. Environmental Protection Agency above is heavily dependent on a carceral system of laws, regulations, and policies. Furthermore, in a country built economically on the backs of Black American enslaved ancestors and the genocide of Indigenous peoples, the dependence on a carceral system is, if nothing else, predictable. In a country where the National Park System was developed, implemented, and enforced to exclude Black and Indigenous peoples based on racism and eugenics, one must ask why climate change solutions created in a carceral state have been unsuccessful in mitigating climate disaster.[31]

To understand environmental justice and climate change we must first understand the continued harm to Black and Indigenous peoples at the hands of systemic oppression. The state government of North Carolina was attempting to dump toxin-laced soil in Warren County; the term "environmental racism" was born out of a speech by Dr. Benjamin Chavis Jr. in 1982 at a protest of the dumping of this soil laced with a carcinogen. However, when we consider the impact of redlining, the Red Summer of 1919, the COVID-19 pandemic, and the continued death of Black bodies at the hands of carceral systems, we understand that environmental racism is a repackaged version of historical racial oppression and settler colonialism.[32]

Contemporary Colonialism: Environmental Racism

Environmental racism is inextricably tied to climate change and its disproportionate impact on Black Americans. When looking at the effects of extreme temperatures, hurricanes and flooding, and wildfires, Black Americans experienced increased chances of mortality, emergency room hospitalization, preterm births, emergency mental health visits, cardiovascular issues, as well as respiratory issues that led to hospitalization.[33] It is not enough to say that Black Americans are no longer enslaved; the United States of America was built off the backs of enslaved peoples, and the impact of that subjugation is still present today. This extends into the natural world and the discussion around climate change. Whiteness as a dominant framework is maintained in local and global environmental movements, normalizing white subjectivity as the only legitimate caretakers of the land.[34] The discourse works to produce environmentalism as a space where white identities safeguard and maintain the land rather than consume or destroy it.[35] This reproduction of whiteness confers the final act of erasure of Native people—white settlers appropriate and perform the dominant notions of Indigenous culture.[36] The erasure of colonization's role in current climate crises is connected to the erasure of what Indigenous peoples experienced as apocalyptic violence during colonialism and continued through today.[37] Until Americans can fully reconcile the history of enslavement and the forced removal of Indigenous peoples from their lands, there will continue to be a disconnect on the best pathway forward to mitigate climate change.

Recent studies have found that colonization led to the abandonment of enough cleared land in the Americas that the resulting terrestrial carbon uptake had a detectable impact on atmospheric carbon dioxide and global surface air temperatures in the two centuries prior to the Industrial Revolution.[38] These findings demonstrate that despite the beliefs that industrialization staged the beginning of climate change, we were already there thanks to colonialism. It was actually the devastation of settler colonialism that led to the current state of climate crises, as many infrastructure systems have been purposely used as tools of dispossession and the extraction of Native lands and labor and the resources of minoritized groups.[39]

The ongoing erasure of Native peoples and history from the land serves to reify the wilderness as an empty space; these narratives continue to entitle white people to occupy and claim originary status, rendering wilderness and environment as a white space.[40] Present-day disparities in ecological amenity access and exposure to hazards are inseparable from long-standing racist patterns of real estate development and associated federal and local policies; for example, the purposeful clearance of Black communities through "urban renewal," exemplified in urban greening—the use of tax increment financing and other incentives to spur new development in poor neighborhoods, encouraging and funding green gentrification.[41] Without analysis of unequal consequences of environmental destruction on all racialized communities, environmental education can become a place where good white people can maintain superiority by saving both the environment and people of color.[42]

A long-standing belief has been that the national parks belong to everyone; in fact this has been a popular slogan of the National Park Service, as well as the Department of Fish and Wildlife. These statements are not only wildly inflammatory given the historical context, but they also negate the equity concern of engaging in human–nature interactions. Black Americans were legally segregated from natural parks and public lands, and even after the legal segregation ended, further legislative decisions made access to green spaces difficult and dangerous. In present-day America due to the criminalization of the Black body, systemic racism, and the carceral state, Black Americans find it difficult to engage in spaces from which they have been historically excluded.

Historians of color have demonstrated the importance of understanding the production of cities through the conjoined evils of capitalism, slavery, and colonialism.[43] Urban greenery has historically been weaponized against people of color and the poor, most recently through aggressive green real estate developments that gentrify portions of the city; these profound inequalities are the defining characteristics of socioecological arrangements within settler colonial states and have been long hidden in plain sight but are slowly becoming part of the public discourse in sustainable design and environmental justice.[44] Ultimately, urban systems hold power to care or destroy within formal institutions, lording the capacity to enact sanctioned violence against whomever it sees fit.

Nature deprivation within the Black community not only comes with health implications, but is also systemically made. The consequence of nature deprivation includes a deterioration in physical and mental health and a reduced acknowledgement of nature, pro-environmental attitudes, and behavior.[45] This loss of human–nature interactions is particularly true for the Black community. More specifically, 68 percent of racially marginalized Americans live in nature-deprived areas, 70 percent of low-income communities are more likely to experience nature deprivation, and in twenty-six states racially marginalized communities experience the highest levels of nature deprivation.[46]

While there are many examples, illustrated by Carolyn Finney, of Black joy within nature, it is not uncommon, as is noted by Kang and colleagues, that Black Americans often prioritize perceived immediate concerns such as community violence and poverty as primary within their communities.[47] This is often misconstrued as Black Americans not caring for the natural world and not loving nature. However, we know from research conducted by Parker and

McDonough that it is not that Black Americans do not care or are not concerned with the natural environment; it is that their concerns look different than their white counterparts.[48]

Continued Erasure

It is not enough to simply conclude through data that Black Americans don't engage with the natural environment nor care about climate change. Research methodologies have always and continue to utilize colorblind approaches to designing, collecting, and analyzing data. Black Americans have deep ties to the land of this country; for centuries enslaved peoples were forced to live, breathe, and die by the land to which they were forcibly brought.[49] Even after emancipation, Black Americans' enslaved ancestors were still beholden to their owners under a new system of oppression, sharecropping, unable to leave farming until all debts were paid; this was ensured under the Black Code, a means to restrict Black bodies and to assert white supremacy. This practice continued after the Great Migration and developed into Jim Crow, all while Black bodies were lynched just for existing. These practices exist today under different names: police brutality, the Flint, Michigan, water crisis, the exponential spread of COVID-19 within the Black community. This is collective trauma for Black Americans and a critical wounding of how they engage with the natural environment. This collective wounding is based in fear and is passed down the generations through epigenetic ties. Dr. Joy DeGruy coined the term "post traumatic slave syndrome" to thoroughly capture the epigenetic experience of Black Americans and how they engage with the world, and within systems of oppression.[50] This is a critical fear of Black Americans engaging within the natural environment.

Historically engaging with the natural environment was and still can be a dangerous experience. If an individual's first response is fear when engaging with the natural environment, it is sufficient to say that the engagement would not continue to occur. Black American ancestors were horrifically traumatized for centuries; this fundamentally changed gene expression, which is still being passed down to our children today. Black Americans have enslaved ancestors, and for generations Black Americans have been living under a system of oppression, racism, and violence. This fact bleeds into our interactions with the world, with our families, the stories we tell, and how we raise our children: always ready for the threat of violence, the overt and covert racism and anti-Blackness, and the systemic implications to our daily lives. This has extended outside of our beings and into our connection with the land, a land that we do not have indigeneity for, because for Black Americans our indigeneity was taken from us in the name of white supremacy and colonialism. It is not that we do not care about the environment, nor that Black people don't engage with the natural environment, nor is it that we are ambivalent to the climate crisis; the response is simply the result of white supremacy.

For generations, white people enslaved and forced our ancestors to engage with the land through horrific violence; the fear is coded into our genetic makeup, and the stories are not far removed so they are living within our community of how the natural environment isn't safe. It is not that "Black people don't do that," it is that white supremacy and systems of oppression have removed our access and our ability to believe the current environmentalism movement is with Black people's best interest and with Black liberation in mind. Bill Gwaltney, a prominent Black environmentalist, summed up the fear in a very blunt statement: "There's a lot of trees in the woods and rope is cheap."[51]

Notions of Justice

Justice is a normative and subjective goal that encapsulates the quality of human relationships with land, improved well-being, dignity, and community membership, and redress for harm.[52] Thinking around justice and nature-based solutions has been dominated by anthropocentric, utilitarian, and economic notions of justice.[53] Ecological justice is meant to address harms, needs, and desired futures while also identifying synergies with human-focused Environmental Justice movements, highlighting harm on living beings composing ecosystems caused by city development.[54] Ecological justice aligns with existing concepts of making space for nature and recognizes the agency of the earth by seeking intersectional and transformational approaches to repair environmental harms.[55] Some put forth climate resilience instead of climate justice; however, the resilience discourse too often forecloses on alternative futures by implicitly invoking a dependence on and return to the logics of racial capitalism; these modes of accumulation are not only contingent upon racial difference for capitalism's continued growth, but also are deeply entrenched in upholding racial disparities.[56]

White Eco-Justice Centers White Solutions

The solutions to the climate movement up until this point have been rooted in settler colonial notions of sustainability: maintaining the status quo. Companies tell their constituents that if you use a reusable water bottle, they'll save the planet. If people would just go vegan, or even vegetarian, that is how climate change will be mitigated. Corporate employers will pay for public transportation passes for their employees, and yet cities will still have air quality ozone layer concerns due to the number of emissions from cars. Legislators are beholden to the political cycle and the constant pendulum swing of a two-party system: one who agrees that climate change is real, and another who denies its existence. Even with legislative power, interventions are focused on industry, clean energy, jobs, and infrastructure changes. While important, these concerns are hardly immediate to the climate change disasters that are happening at this very second.

Eco-justice centered in whiteness, centered in white solutions, will not help mitigate climate disaster. Rugged individualism, climate anxiety, and apocalyptic doomsday plans will not save us from climate-related disasters or health concerns. White Americans are so concerned with the advancement of Black and Indigenous peoples at their expense that they refuse to look at the systems in place that also impact how they exist within the world.[57] This rugged individualism extends into environmental justice; it is how we have handled the climate change crisis, and it is grounded in white supremacy—and not effective. Imagining better for the world requires new ideas, new thought processes, and dismantling systems that were designed to harm. The only way through climate disasters is collective action. Collective action has been the root of Black and Indigenous people's survival, working together within community to solve problems. Collective action doesn't remove the space for individualistic action, quite the opposite; the individual is part of the collective, and the collective and the individual are serving a common goal.

This was a primary highlight of the COVID-19 pandemic. Many people collectively made masks and donated old masks and personal protective equipment to hospitals and medical personnel. In New York City the community began to sing to health-care workers as a means for solidarity.[58] Black and Indigenous people are not only disproportionally impacted by climate change, they are also disproportionately impacted by the climate disasters that occur. The experience of climate anxiety by white Americans versus their Black and Indigenous counterparts is different. White Americans rely on the colonial and systemic systems they built to oppress Black and Indigenous people to save themselves from climate disaster and climate change, whereas Black and Indigenous peoples fundamentally understand that those systems were not made to support nor save them from climate change and climate disaster.

Black and Indigenous people put the power within their community, and within helping each person out as a collective resistance. It is how Black and Indigenous people engage with the natural environment, as a relationship and a collective honoring of the environment. We see this with the Black Panther Party, and the Black Church and their advocacy for climate change mitigation. While white people utilize the living areas of Black and Indigenous people and pollute them with toxic waste, landfills, and industrial pollution, Black and Indigenous peoples work together to provide resources for each other to ensure they are covered.

Sustainability Is Not the Goal

Sustainability has no place in a decolonial liberation-based world. To sustain means to maintain. If maintaining current climate mitigation efforts were working, there would be no need for collective action, nor legislative action. Maintaining the status quo is not serving the Environmental Justice movement; it is not protecting Black and Indigenous people, which is hardly a surprise given that maintaining the status quo maintains systems of oppression and as such is not made with Black liberation and Indigenous sovereignty in mind.

To effectively mitigate climate change and build better climate futures there must be a focus on regenerative climate solutions. Regenerative climate solutions upend the status quo as they no longer rely on rugged individualism; they are global, national, state, and local issues. It is no longer about survival of the fittest; it is moving toward a collective being. However, considering the current systems in place are rooted in white supremacy, it is hardly a surprise that this has not been achieved. White American culture refuses to acknowledge their historical wrongdoings and continues to subjugate Black and Indigenous bodies through the systems they created. Through this refusal, the regenerative climate futures cannot exist; it is the antithesis of regeneration to maintain the status quo, but until white America opts to learn from Black individuals on how to develop and maintain community and resilience, until they return Indigenous sovereign land to the rightful nation's sustainability, it will be the only achievable methodology, a methodology that is still causing irreparable damage to Black and Indigenous bodies, and to the land.

Decolonize Environmental Justice

As we work toward sovereign and liberated climate futures, proposed solutions must be enacted with sensitivity to local context and history and consider whose goals are to be met through these solutions, whose knowledge and power shapes these future programs and policies.[59] Indigenous people must be at the front of Environmental Justice movements, as we have demonstrated since time immemorial that our struggle for social justice is intrinsically tied to ecological justice. Indigenous scholars confirm the need to understand ecological crises as an intensification of colonialism; thus decolonization is necessary for viable sustainable paths forward to be envisioned.[60] Maintaining ecological resilience locally and globally requires sustaining and defending and returning Indigenous lands.[61]

As colonialism has impacted all of us, it is the collective responsibility for each and every one of us to decolonize.[62] Decolonization is the undoing of colonialism, a process, a theory of change, and a framework for envisioning and achieving the futures we want to see for the next generations. Decolonizing environmental justice illuminates these historic injustices emanating from settler colonialism and manifesting as present-day environmental racism and white supremacy.

To further these aims, abolition emerges as a necessary counterpart. Abolition is a theory of change, a way of understanding how Indigenous life and politics have been shaped by colonialism without being defined by it, but is not just about a vision, it is about our everyday practice, recognizing that carceral spaces such as incarceration, policing, and deportation are not the exception, but rather the everyday spaces where people of color struggle to live their lives.[63]

Abolition and decolonization both aim to dismantle the systems that seek to oppress and perpetuate harm; there can be no horizon for justice or successful anticolonial struggle for Indigenous communities within the existing structures of global capitalism.[64] Abolition teaches about both rooting and reaching, grounding the work in place-based ways of knowing and learning.[65] Abolition teaches us that the prison nation governs through erasure and abandonment, but that prison is everywhere; therefore, abolition must be everywhere too.[66] Similarly, decolonization reminds us to pay attention and be accountable to how coloniality plays out in the places where we live and work.[67] Holding these marginal, liminal spaces as decolonial abolitionists subverts the hegemony of white supremacy.[68] As we have previously introduced, the present work sets the stage for questioning the very foundations of eco-justice scholarship and invites us to truly reckon with what it would look like to ground solutions in Indigenous cosmologies and abolition ecologies.

Indigenous Cosmologies

Decolonial approaches to environmental injustices would greatly benefit from understanding how Indigenous peoples conceptualize the origin and development of the universe. Decolonization makes space for multiple ways of knowing and being, with the revitalization of Indigenous ways of knowing and being with land as central to addressing climate change.[69]

Western environmental governance approaches emphasize utility and responsibility to country, in alignment with capitalist and individualistic notions; however, Indigenous environmental knowledge understands ecological systems as societies with ethical interspecies structures, the land as a sentient and feeling being, and that natural elements are seen as kin in reciprocal relationships, centering the importance of positionality, situatedness, and responsibility for stewarding said knowledge.[70] Indigenous communities view this as a community responsibility; this is our way of being a good relative both with the land and one another and, most importantly, for our future generations. A relational framework is a decolonial practice that further highlights the ways colonization has set the stage for industrialization to cause climate change.[71] This is an embodied worldview and way of being, not merely an academic framework that can be understood in a Western institution.

Indigenous communities have enacted modes of survival and resistance that preserved our ways of knowing and being that did not rely upon hierarchies between animate and inanimate, maintaining our own autonomous definition of self, collectivity, and culture, including plant, animal, and geographic kin as ancestors, knowledge keepers, and essential parts of their definition of community, self, and life. This cultural resilience survived through passing down memories of medicine making, tending to the land despite cultural repression and the routine violence of colonialism.[72] Resisting hierarchies and reclaiming our traditional practices are both instances of embodied decolonial praxis.

Indigenous communities subvert the dominant discourse—our existence is resistance. While Native Nations are particularly vulnerable to the impacts of climate change, we also possess generations of experience from survival, thus equipping Indigenous communities to survive catastrophic environmental change; Native peoples have adapted to and survived through these changes for centuries using our own knowledge.[73] We are the first scientists. Thus, the exclusion of our experiences and contributions is another form of environmental injustice.

Climate scholars can decolonize their Environmental Justice practice by embodying grounded normativity in their scholarship. Grounded normativity is the need for explicit ethical, multigenerational relationships with place in order to achieve positive ecological transformation; recognizing reciprocity is required for healthy human–ecological relationships, caring for the land as the land cares for us.[74] However, reciprocity is not a transaction or tit for tat, but a practice and worldview, a way of being. Grounded normativity also posits that we must accept and interact with the land as a living entity as it helps address the relationship between social and ecological justice as a practice of solidarity across differences yet still depends on specific, Native place-based practices and forms of knowledge.[75]

Indigenous Environmental Justice

Current frames of environmental justice do not account for the complexity of Indigenous intergenerational environmental justice. Theoretical discussions of environmental justice must recognize tribal sovereignty, cultures, and identities through Indigenous ontologies and epistemologies rather than through Western liberal thought and governance approaches.[76] From an Indigenous view, environmental injustice, including the climate crisis, is inevitably tied to and symptomatic of ongoing processes of colonialism, dispossession, capitalism, imperialism, and patriarchy; thus, applying settler colonialism and environmental injustice analyses reveals deep-rooted historical forces within every facet of injustice experienced by Indigenous communities.[77] In order to advance and achieve social-ecological justice, it is vital to develop understanding and reinstate Indigenous approaches to ecological governance and stewardship. For example, a crucial aspect of the current climate crises surrounds the violation of respect for nonhuman life.[78] Indigenous environmental justice applies beyond the human dimension, as these animacy hierarchies relegate all nonhuman beings, like animals, plants, and geographic life, as less than human, based on perceptions of them as merely inanimate matter. Ultimately, humanity alone does not have the solutions to save us from ourselves; Indigenous environmental justice includes all relations.[79]

Indigenous environmental justice acknowledges the purposeful elimination of Indigenous ecological knowledge and works toward its restoration as Indigenous peoples remain connected to the land despite removal and displacement from our ancestral homelands. Some scholars report that Indigenous environmental justice is not about romanticizing Indigenous environmental governance or dismissing Western "scientific" approaches, but about recognizing how historic legacies continue to shape social decision-making about the environment in ways that impact what is deemed important, desirable, and just. Yet, even in their description of Indigenous environmental justice, the authors demonstrate their limited perspective of these matters as they pit Indigenous knowledge against Western "science." However, traditional Indigenous knowledge is science, and the separation of Indigenous knowledge and science is a false dichotomy. Indigenous communities are the first scientists of this land and have stewarded it since time immemorial. Indigenous environmental justice restores our traditional systems of governance and knowledge while also making space for diversity of social-ecological practices of various marginalized communities.[80]

Indigenous environmental justice focuses attention on dismantling systems of power that degrade human–land relationships as compared to seeking an extension of rights by the same colonial system.[81] Environmental Justice embodies decolonization by recognizing and practicing our own traditions and knowledge rather than band-aid solutions that leave these harmful settler systems intact.

Abolition Ecologies

Abolition ecology builds on direct action traditions that began in the abolitionist movement against slavery, were core tactics in the civil rights movement, and continued through today with the Black Lives Matter movement, as human history has shown that rights are seldom granted to marginalized communities, but rather won through struggle. Abolition ecologies highlight the key role white supremacy plays in shaping nature–society relations, recognizing the deeper racialized ways that nature has been unevenly produced.[82] Abolition ecology pulls together participatory discourses from the Black radical tradition and political ecology to strive for a more liberatory land-based ethic that challenges the oppressive logics of racial capitalism.[83] Abolition ecologies embody the Black radical tradition of freedom dreams shared across differences, moving beyond a Black/white binary, and puts forth a nuanced critique of white supremacy through land struggles amid abolition and decolonization.[84] Abolitionist principles center historic racisms and how intersectional drivers of trauma can be incorporated into climate adaptation responses.[85] Abolitionist ecologists purport that relearning lessons that have been lost or erased from history is one of the most powerful tools for building solidarity toward liberation.[86]

Abolition ecologies understand ways that reciprocal land relationships are often synonymous with liberation struggles, as the goal of abolition ecology is theorizing and organizing around the white supremacist logics that perpetuate uneven racial development in land and property relations.[87] Our struggles for sovereignty and liberation are inextricably linked; thus an understanding of both Indigenous cosmologies and abolition ecologies is necessary to move toward decolonizing environmental justice. Abolition ecology invites us to think about how to build freedom across relationships of land and people; when viewed through the lens of liberation, both human and nonhuman relationships look different; for example, enslaved individuals who self-emancipated through difficult journeys experienced terror in the landscape. After self-emancipation they would hide in the wilds, Indian Country, and other spaces inaccessible to white settler states in order to protect their lives; many communities were made through the solidarity of Native peoples accepting self-emancipated formerly enslaved people into their homelands.[88]

Critical environmental justice and Indigenous environmental justice both focus on how governance cannot be relied on to merely extend the rights of nature or marginalized communities, much less appropriate reparations or return land; thus, we must shift our attention to transforming the systems that caused harm in the first place.[89] We can utilize short-term and long-term solutions, transforming existing institutions and building up new institutions focused on land-based justice to make freedom as a place.[90] Abolition ecologies employ blockades in a twofold manner, first aspiring to dismantle the system of climate adaptation practices that work to re-create systemic violence under the guise of resilience, and second, building in its place new institutions and solidarity that can sustain abolitionist visions for the future. Abolitionists seek liberated lifeways through a commitment to radical place-making, seeking to liberate the environment through envisioning freedom as a place. Ruth Wilson Gilmore's notion of "freedom as a place" describes efforts to dismantle racist institutions and build coalitional land-based politics with place-making potential.[91] Abolition ecology demands attention to the ways that coalitional land-based politics dismantle oppressive institutions and challenges the general notion that territory is alienable and exclusive, attending to the reciprocal relationships between Indigenous folx and land.[92] A geography of abolition must center understanding of territory as a Native Nation's homeland.

Rather than define communities by the violence they have suffered, abolition ecologies call for us to pay attention to the radical place-making and land-, air-, and water-based environment where places are made; imagine if we defined communities by the places they have made for themselves, rather than by the shared violence of colonialism.[93] Place-making as a process of building institutions and processes focused on political imperatives of access to fresh air and clean water is a vital preconfiguration for liberation.[94]

Conclusion

We must imagine new futures and think beyond the bounds of colonial logic. Abolitionist climate justice praxis necessitates seeing beyond physical inundation, dismantling racial regimes of ownership that are imbued in property relations. Even the language we use matters, such as distinguishing between "property" and "land," problematizing colonial logics of property compared to land as a relative, full of life and stories. Embodied knowledge is a site for radical activism, seen in both Indigenous cosmologies and abolition ecologies; decolonization reminds us that knowing is not the same as doing, while abolition ecologies are steeped in the abolitionist principles of agitation and action, working to produce nature that bolsters the cultural importance of Black life as opposed to vanquishing it.[95] We must mobilize the historic knowledge that has been stripped from people and land.

Achieving liberation requires abolishing the ongoing systemic processes of hierarchy, dispossession, and exclusion; in order to limit environmental racism, we must recognize the racialized production of uneven capital accumulation by environmental dispossession.[96] We must invest in the process of recognizing how treaty rights and lack of have affected Indigenous peoples and the social-ecological system as a necessary step to justice as there are far more deep histories of nature that are produced through settler colonialism and racial capitalism than the field has grappled with.[97]

White settler fears of climate change prepare for an end of capitalism and Western civilization, while Two Spirit visions plant seeds for a future that is populated by overgrown, wild, unruly bodies, landscape, and sexualities. Indigenous communities have faced the end of the world many times before and have more than survived. We will continue to fight for our ancestral lands, plants, and animal kin. As queer Black and Indigenous scholars, we hold firm in the abolitionist practice of refusal to disappear or be disappeared. We are still here.

NOTES

1. Zbigniew Jakub Grabowski et al., "How Deep Does Justice Go? Addressing Ecological, Indigenous, and Infrastructural Justice through Nature-Based Solutions in New York City," *Environmental Science & Policy* 138 (2022): 172.
2. Grabowski et al., "How Deep Does Justice Go?," 174.
3. Nik Heynen and Megan Ybarra, "On Abolition Ecologies and Making 'Freedom as a Place,'" *Antipode* 53, no. 1 (January 2021): 9.
4. Sheelah McLean, "The Whiteness of Green: Racialization and Environmental Education," *Canadian Geographer* 57, no. 3 (Autumn 2013): 354; Nik Heynen, "Toward an Abolition Ecology," *Abolition: A Journal of Insurgent Politics* 1 (2018): 243.
5. Grabowski et al., "How Deep Does Justice Go?," 175.

6. Meg Parsons, Karen Fisher, and Roa Petra Crease, "Environmental Justice and Indigenous Environmental Justice," in *Decolonising Blue Spaces in the Anthropocene* (Cham: Springer, 2021): 39–73.

7. Heynen and Ybarra, "On Abolition Ecologies," 5.

8. Brooklyn Leonhardt, "Ancestral Lands and Genders: A Queer Indigenous Critique of Settler Climate Change and Post-Apocalyptic Narratives," *Radical Philosophy Review* 26, no. 1 (2023): 21.

9. Jeremy Braithwaite, "Colonized Silence: Confronting the Colonial Link in Rural Alaska Native Survivors' Non-Disclosure of Child Sexual Abuse," *Journal of Child Sexual Abuse* 27, no. 6 (2018): 589.

10. Daniel L. Dickerson and Carrie L. Johnson, "Design of a Behavioral Health Program for Urban American Indian/Alaska Native Youths: A Community Informed Approach," *Journal of Psychoactive Drugs* 43, no. 4 (2011): 337–42.

11. Dickerson and Johnson, "Design of a Behavioral Health Program," 338.

12. Talia Nelson, "Historical and Contemporary American Indian Injustices: The Ensuing Psychological Effects," 2011, Commonwealth Honors College Theses and Projects, Paper 6, https://scholarworks.umass.edu/chc_theses/6; Les B. Whitbeck et al., "Discrimination, Historical Loss and Enculturation: Culturally Specific Risk and Resiliency Factors for Alcohol Abuse among American Indians," *Journal of Studies on Alcohol and Drugs* 65, no. 4 (2004): 409; Braithwaite, "Colonized Silence," 589.

13. Ashish A. Vaidya, "Shadows of Colonialism: Structural Violence, Development and Adivasi Rights in Post-Colonial Madhya Pradesh," *South Asia* 41, no. 2 (2018): 315–30.

14. Grabowski et al., "How Deep Does Justice Go?," 174.

15. Braithwaite, "Colonized Silence," 591.

16. Grabowski et al., "How Deep Does Justice Go?," 175.

17. Gina Kruse et al., "The Indian Health Service and American Indian/Alaska Native Health Outcomes," *Annual Review of Public Health* 43 (2022): 559–76.

18. Grabowski et al., "How Deep Does Justice Go?," 175.

19. Grabowski et al., "How Deep Does Justice Go?," 172.

20. Heather Norris, "Colonialism and the Rupturing of Indigenous Worldviews of Impairment and Relational Interdependence: A Beginning Dialogue towards Reclamation and Social Transformation," *Critical Disability Discourses* 6 (2014), https://cdd.journals.yorku.ca/index.php/cdd/article/view/39665.

21. Leonhardt, "Ancestral Lands and Genders," 21.

22. Heynen, "Toward an Abolition Ecology," 242.

23. Dean Hardy, Maurice Bailey, and Nik Heynen, "'We're Still Here': An Abolition Ecology Blockade of Double Dispossession of Gullah/Geechee Land," *Annals of the American Association of Geographers* 112, no. 3 (2022): 869.

24. Heynen, "Toward an Abolition Ecology," 244; Leonhardt, "Ancestral Lands and Genders," 27.

25. Heynen, "Toward an Abolition Ecology," 245.

26. Grabowski et al., "How Deep Does Justice Go?," 175.

27. Heynen, "Toward an Abolition Ecology," 245.

28. McLean, "The Whiteness of Green," 359.

29. Heidi Kiiwetinepinesiik Stark, "(Re) Mapping Worlds: An Indigenous (Studies) Perspective on the Potential for Abolitionist and Decolonial Futures," *American Quarterly* 75, no. 4 (2023): 847–57.

30. "Environmental Justice," U.S. Environmental Protection Agency, September 6, 2023, https://www.epa.gov/environmentaljustice.

31. Jedediah Purdy, "Environmentalism's Racist History," *The New Yorker*, August 13, 2015.

32. Jennifer D. Roberts et al., "'I Can't Breathe': Examining the Legacy of American Racism on Determinants of Health and the Ongoing Pursuit of Environmental Justice," *Current Environmental Health Reports* 9, no. 2 (2022): 211–27.

33. Alique G. Berberian, David J. X. Gonzalez, and Lara J. Cushing, "Racial Disparities in Climate Change-Related Health Effects in the United States," *Current Environmental Health Reports* 9, no. 3 (2022): 451–64.

34. Eve Tuck and Rubén A. Gaztambide-Fernández, "Curriculum, Replacement, and Settler Futurity," *Journal of Curriculum Theorizing* 29, no. 1 (2013): 72–89.

35. McLean, "The Whiteness of Green," 360.

36. Patrick Wolfe, "Settler Colonialism and the Elimination of the Native," *Journal of Genocide Research* 8, no. 4 (2006): 387–409.

37. Leonhardt, "Ancestral Lands and Genders," 27.

38. Leonhardt, "Ancestral Lands and Genders," 22.

39. Grabowski et al., "How Deep Does Justice Go?," 176.

40. McLean, "The Whiteness of Green," 354.

41. Grabowski et al., "How Deep Does Justice Go?," 176.

42. McLean, "The Whiteness of Green," 358.

43. Heynen and Ybarra, "On Abolition Ecologies," 5.

44. Grabowski et al., "How Deep Does Justice Go?," 175.

45. Masashi Soga and Kevin J. Gaston, "Extinction of Experience: The Loss of Human–Nature

Interactions," *Frontiers in Ecology and the Environment* 14, no. 2 (2016): 94–101.

46. Jenny Rowland-Shea et al., "The Nature Gap," Center for American Progress, July 21, 2020, https://www.americanprogress.org/article/the-nature-gap/.

47. Carolyn Finney, *Black Faces, White Spaces: Reimagining the Relationship of African Americans to the Great Outdoors* (Chapel Hill: University of North Carolina Press, 2014); Joonmo Kang, Vanessa D. Fabbre, and Christine C. Ekenga, "'Let's Talk about the Real Issue': Localized Perceptions of Environment and Implications for Ecosocial Work Practice," *Journal of Community Practice* 27, no. 3–4 (2019): 317–33.

48. Julia Dawn Parker and Maureen H. McDonough, "Environmentalism of African Americans: An Analysis of the Subculture and Barriers Theories," *Environment and Behavior* 31, no. 2 (1999): 155–77.

49. Dianne D. Glave, *Rooted in the Earth: Reclaiming the African American Environmental Heritage* (Chicago: Lawrence Hill Books, 2010).

50. Joy DeGruy, *Post Traumatic Slave Syndrome: America's Legacy of Enduring Injury and Healing* (Portland, OR: Joy DeGruy Publications, 2005).

51. Quoted in Finney, *Black Faces, White Spaces*, 118.

52. Carol A. Hand, Judith Hankes, and Toni House, "Restorative Justice: The Indigenous Justice System," *Contemporary Justice Review* 15, no. 4 (2012): 449–67.

53. Grabowski et al., "How Deep Does Justice Go?," 172; Hand, Hankes, and House, "Restorative Justice," 458.

54. Grabowski et al., "How Deep Does Justice Go?," 172.

55. Grabowski et al., "How Deep Does Justice Go?," 173.

56. Hardy, Bailey, and Heynen, "'We're Still Here,'" 868.

57. Marlene F. Watson et al., "COVID-19 Interconnectedness: Health Inequity, the Climate Crisis, and Collective Trauma," *Family Process* 59, no. 3 (2020): 832–46.

58. Peter Marks, "The Nightly Ovation for Hospital Workers May Be New York's Greatest Performance," *Washington Post*, April 6, 2020.

59. Grabowski et al., "How Deep Does Justice Go?," 171.

60. Deborah McGregor, Steven Whitaker, and Mahisha Sritharan, "Indigenous Environmental Justice and Sustainability," *Current Opinion in Environmental Sustainability* 43 (2020): 35–40.

61. Grabowski et al., "How Deep Does Justice Go?," 172.

62. John T. Gagnon and Maria Novotny, "Revisiting Research as Care: A Call to Decolonize Narratives of Trauma," *Rhetoric Review* 39, no. 4 (2020): 486–501.

63. Heynen and Ybarra, "On Abolition Ecologies," 6 and 2.

64. Michael J. Viola et al., "Introduction to Solidarities of Nonalignment: Abolition, Decolonization, and Anticapitalism," *Critical Ethnic Studies* 5, no. 1–2 (2019): 5–20.

65. Devin Burns et al., "The Ground on Which We Stand: Making Abolition," *Journal for the Anthropology of North America* 23, no. 2 (2020): 98–120.

66. Burns et al., "The Ground on Which We Stand," 100.

67. Hokulani K. Aikau et al., "Indigenous Feminisms Roundtable," *Frontiers: A Journal of Women Studies* 36, no. 3 (2015): 84–106.

68. Burns et al., "The Ground on Which We Stand," 117.

69. Leonhardt, "Ancestral Lands and Genders," 21.

70. Grabowski et al., "How Deep Does Justice Go?," 174.

71. Leonhardt, "Ancestral Lands and Genders," 27.

72. Leonhardt, "Ancestral Lands and Genders," 25.

73. McGregor et al., "Indigenous Environmental Justice," 37.

74. Grabowski et al., "How Deep Does Justice Go?," 175.

75. Heynen and Ybarra, "On Abolition Ecologies," 6.

76. Parsons, Fisher, and Crease, "Environmental Justice and Indigenous Environmental Justice," 39.

77. Grabowski et al., "How Deep Does Justice Go?," 37.

78. Leonhardt, "Ancestral Lands and Genders," 27.

79. McGregor et al., "Indigenous Environmental Justice," 36.

80. Grabowski et al., "How Deep Does Justice Go?," 171.

81. Grabowski et al., "How Deep Does Justice Go?," 172.

82. Heynen and Ybarra, "On Abolition Ecologies," 7.

83. Hardy, Bailey, and Heynen, "'We're Still Here,'" 868.

84. Heynen and Ybarra, "On Abolition Ecologies," 7.

85. Hardy, Bailey, and Heynen, "'We're Still Here,'" 868.

86. Heynen and Ybarra, "On Abolition Ecologies," 3.

87. Leonhardt, "Ancestral Lands and Genders," 21.

88. Heynen and Ybarra, "On Abolition Ecologies," 7.

89. Grabowski et al., "How Deep Does Justice Go?," 174.

90. Heynen and Ybarra, "On Abolition Ecologies," 9.

91. Hardy, Bailey, and Heynen, "'We're Still Here,'" 868. Ruth Wilson Gilmore, *Abolition geography: Essays towards liberation* (New York: Verso Books, 2022).

92. McLean, "The Whiteness of Green," 354.

93. Heynen and Ybarra, "On Abolition Ecologies," 3.

94. Robyn Moran and Lisbeth A. Berbary, "Placemaking as Unmaking: Settler Colonialism, Gentrification, and the Myth of 'Revitalized' Urban Spaces," in *Leisure Myths and Mythmaking*, ed. Brett Lashua, Simon Baker, and Troy D. Glover (London: Routledge, 2022), 106–20.

95. Hardy, Bailey, and Heynen, "'We're Still Here,'" 868.

96. Hardy, Bailey, and Heynen, "'We're Still Here,'" 869.

97. Grabowski et al., "How Deep Does Justice Go?," 176; Heynen and Ybarra, "On Abolition Ecologies," 2.

Conclusion

Decolonizing Environmental Justice Pedagogy

Tatiana Konrad

———————————————————————————————————

Bringing together cross-disciplinary scholars and connecting the issues of environment, climate, health, race, and justice, this book attempted to uncover the often overlooked or neglected relationships that are at the core of environmental justice. Drawing on geographies and approaches from both the Global North and Global South, this book revealed the various manifestations of environmental injustice as well as the consequences thereof. Emphasizing the global nature of the problem, however, the contributors demonstrated how race remains the key factor that guides the uneven distribution of environmental toxicity and unhealth more broadly, focusing on the cases of environmental injustice that oftentimes remain at the community level, rather than rising to a national, let alone global, concern. As the book has shown, while the local examples of environmental injustice might vary, the reason for it, namely power relations, structures, and institutions that cause and sustain these kinds of injustice, is very similar. These are the ramifications of colonialism that subjugated peoples and nature; and while colonialism as a political system is over, the ideology that it produced continues to be embedded into decision-making processes related to environmental, climate, and health crises, generating inequality worldwide.

Unpacking these serious issues and working at the intersection of multiple disciplines, *Race and Environmental Justice in the Era of Climate Change and COVID-19* contributes to the existing bodies of knowledge and ongoing research on environmental justice. Nevertheless, this book can only do some of the necessary work. While we are proud to help advance the important scholarly and activist work on justice, we also recognize that much is left to be accomplished. Intersectionality is key when addressing the issue of justice, and specifically in the context of scholarly research, we recommend that investigative teams should be diverse, both to better understand the problem and to facilitate inclusion. For example, research on disability in relation to the core issues probed in this book would uncover important new perspectives on health and justice. Exploring the injustices experienced by people who find themselves on multiple edges of disadvantage, like members of BIPOC communities with disabilities, will allow us to more fully comprehend the multiple ways in which injustice operates and penetrates people's lives. Furthermore, it will produce a better understanding of the type of activist work that needs to be done and changes that must be implemented on sociopolitical, institutional, economic, and cultural levels.

Incorporating environmental justice in curricula is important scholarly activist work that educators can and should do. Solving the ongoing environmental crisis is impossible without possessing sufficient knowledge: how and which anthropogenic activities transform and destroy the planet and planetary health, whom these actions affect the most, what measures might be taken to prevent environmental inequality, and why injustice persists. Environmental education is one step toward a solution to the environmental crisis, and situating environmental justice at the center of that pedagogy is crucial. This book is one such instrument for teaching environmental justice at universities and high schools. Though it is difficult to give any specific recommendations on how to best incorporate the book in a given educational setting, we are convinced it can provide solid materials for courses in the environmental humanities, anthropology, and cultural geography, as well as those that focus broadly on literature and the environment, globalization, race, and social justice.

Research on decolonial environmental justice, including how to teach it, is ongoing.[1] And this is perhaps the most crucial work that the scholarly activist community has been doing to promote environmental justice: unpicking the complex, intertwined issues, making calamities and injustices visible, including voices from BIPOC communities, and making sure that decision-making practices become inclusive worldwide. Theorizing—and also implementing—environmental justice on a daily basis, in every practice, action, and decision, is the only way to consciously fight for a better future for all. And we persevere.

NOTE

1. Some helpful sources on teaching decolonial environmental justice include Karen Jarratt-Snider and Marianne O. Nielsen, eds., *Indigenous Environmental Justice* (Tucson: University of Arizona Press, 2020); Malcom Ferdinand, *Decolonial Ecology: Thinking from the Caribbean World*, trans. Anthony Paul Smith (Hoboken: Wiley, 2021); Jan Wilkens and Alvine R. C. Datchoua-Tirvaudey, "Researching Climate Justice: A Decolonial Approach to Global Climate Governance," *International Affairs* 98, no. 1 (2022): 125–43; Sikina Jinnah et al., eds., *Teaching Environmental Justice: Practices to Engage Students and Build Community* (Cheltenham: Edward Elgar Publishing, 2023); Michael Simpson and Alejandra Pizarro Choy, "Building Decolonial Climate Justice Movements: Four Tensions," *Dialogues in Human Geography*, May 8, 2023, 1–4; Mariana Walter, Lena Weber, and Leah Temper, "The EJAtlas: An Unexpected Pedagogical Tool to Teach and Learn about Environmental Social Sciences," in *The Barcelona School of Ecological Economics and Political Ecology: A Companion in Honour of Joan Martinez-Alier*, ed. Sergio Villamayor-Tomas and Roldan Muradian (Cham: Springer, 2023), 211–18.

Contributors

Autumn Asher BlackDeer is a queer decolonial scholar-activist from the Southern Cheyenne Nation and serves as an assistant professor in the Graduate School of Social Work at the University of Denver. Her scholarship illuminates the impact of structural violence on American Indian and Alaska Native communities. BlackDeer centers Indigenous voices throughout her research by using quantitative approaches and big data as tools for responsible storytelling. BlackDeer is a racial equity scholar with an emphasis on Indigenous tribal sovereignty and is deeply committed to furthering decolonial and abolitionist work.

Helen Bond is a university professor in the Department of Curriculum and Instruction in the School of Education at Howard University in Washington, D.C. She is also the former director of the Center for Excellence in Teaching, Learning, and Assessment. Bond is a Fulbright-Nehru Scholar to India, cochair of the United Nation's Sustainable Development Solutions Network (UN SDSN), and executive council member to the Center for Women, Gender, & Global Leadership at Howard University. Bond is also the faculty liaison to the Center for African Studies at Howard University. Howard is one of only ten

U.S. universities and the only historically Black college and university (HBCU) designated by the U.S. Department of Education as a comprehensive National Resource Center for African Studies. With a PhD in human development, Bond's expertise is in teacher education, education for sustainable development, and race and human development.

Tagaaq Evaluardjuk-Palmer was born near Igloolik and lived near Pond Inlet on a traditional camp; as a child she grew up in Iqaluit. She first moved to Manitoba in 1980 and has lived in Churchill, Thompson, Winnipeg, and Brandon. Since moving to Manitoba, she has volunteered in different capacities, for example, addictions recovery and self-help groups in Thompson (1988–2002), Midwifery Council of Manitoba (1999–2001), and as an Inuit knowledge keeper on Brandon University's Indigenous Education Senate Sub-Committee (2018–19). Tagaaq is now a member of the Inuit Elders Executive Council Qanuinngitsiarutiksait Project with Ongomiizwin—Indigenous Institute of Health and Healing at the University of Manitoba; since 2020 she has been part of the National Centre for Truth and Reconciliation and the Residential School Survivor's Circle.

Juan M. Floyd-Thomas is associate professor of African American religious history in the Divinity School and Graduate Department of Religion and Affiliated Faculty in Religious Studies at Vanderbilt University in Nashville, Tennessee. In addition to several journal articles, book chapters, and other publications, Floyd-Thomas is author of *The Origins of Black Humanism: Reverend Ethelred Brown and the Unitarian Church* (2008) and *Liberating Black Church History: Making It Plain* (2014) as well as coauthor of *Black Church Studies: An Introduction* (2007) and *The Altars Where We Worship: The Religious Significance of Popular Culture* (2016) and coeditor of a volume titled *Religion in the Age of Obama* (2018). He is currently working on two book projects: a history of African diasporic religions in Harlem, and a collection of essays on white supremacy, Christian nationalism, and the culture wars.

Bishnupriya Ghosh teaches in global media at the University of California, Santa Barbara. She has published two monographs on the cultures of globalization: *When Borne Across: Cosmopolitics in the Indian Novel* (2004) and *Global Icons: Apertures to the Popular* (2011). Her current research is on media, risk, and globalization: the coedited *Routledge Companion to Media and Risk* (2020) and a new monograph on viral pandemics, *The Virus Touch: Theorizing Epidemic Media* (2023). She is starting research on media environments of viral infection in a book of essays tentatively titled *Epidemic Intensities*.

Tatiana Konrad is the principal investigator of "Air and Environmental Health in the (Post-)COVID-19 World," a postdoctoral researcher in the Department of English and American Studies, University of Vienna, Austria, and the editor of the "Environment, Health, and Well-being" book series at Michigan State University Press. She holds a PhD in American studies from the University of Marburg, Germany. She was a visiting fellow at the University of Chicago (2022), a visiting researcher at the Forest History Society (2019), an Ebeling fellow at the American Antiquarian Society (2018), and a visiting scholar at the University of South Alabama (2016). She is the author of *Climate Change Fiction and Ecocultural Crisis: The Industrial Revolution to the Present* (2024) and *Docu-Fictions of War: U.S. Interventionism in Film and Literature* (2019); the editor of *Disability, the Environment, and Colonialism* (2024), *Imagining Air: Cultural Axiology and the Politics of Invisibility* (2023), *Plastics, Environment, Culture, and the Politics of Waste* (2023), *Cold War II: Hollywood's Renewed Obsession with Russia* (2020), and *Transportation and the Culture of Climate Change: Accelerating Ride to Global Crisis* (2020); and a coeditor of *Cultures of War in Graphic Novels: Violence, Trauma, and Memory* (2018).

Josée G. Lavoie is a professor of community health sciences at Ongomiizwin—Indigenous Institute of Health and Healing at the University of Manitoba. Josée is an established academic working with Indigenous communities in Canada and Australia. The overall goal of her research is to improve access to health services in rural, remote, and Indigenous environments, by undertaking broad-based research that documents the links and disconnects that exist between national, provincial/state, and Indigenous health policy and planning, identifies gaps, and proposes solutions. She currently leads a study in partnership with the Manitoba Inuit Association, and coled by Elders from the Kivalliq region of Nunavut.

Guillermo López Varela is a research professor at the Intercultural University of the State of Puebla and is currently developing participatory action research projects in the Ngigua region of Puebla, Mexico. He is coordinator of the working group "Food from the Americas" of the Latin American Sociological Association and is a member of the academic group "Pueblos originarios, su bienestar físico, mental y cultural a lo largo de la vida" (BUAP VIEP-22023). He has published articles, books, and book chapters in Canada, the United States, Colombia, Greece, Guatemala, Mexico, and Argentina. He is a member of the National System of Researchers and registry of the Mexican Ministry of Public Education.

María Cristina Manzano-Munguía is a research professor at the Institute of Social Sciences and Humanities at the Benemérita Universidad Autónoma de Puebla in Puebla, Mexico and adjunct professor in the Department of Canadian and Indigenous Studies at Mount Allison University (PhD, University of Western Ontario, London, Ontario; MA, University of Guelph, Ontario; BA, University of the Americas-Puebla, Puebla, Mexico). Member of the National Research Council since 2014, she belongs to the academic research group "Racism, Identities, and Modes of Subjectivity." She has published on issues related to Indigenous forced transnationalism across borderlands, First Nations transnationalism, Canadian Indian diaspora, Indian policy and legislation in Canada, Indigenous mobility, and Indigenous return migration among others. Her current research interests include Indigenous transnationalism, return migrant students' citizenship and participation, and Indigenous intercultural education. She was the recipient of the "Phillips Fund for Native American Research" from the American Philosophical Society in 2012 and the "Democracy, Diasporas, and Canadian Security in International Perspective" in 2007–8, at the York Centre for International and Security Studies, York University, Toronto, Ontario. She is a fellow of the Salzburg Seminar American Studies Association (SSASA) and of the American Philosophical Society.

Sierra Roach Coye is a queer Black scholar-activist from Haudenosaunee territory and is a PhD student in the Graduate School of Social Work at the University of Denver. Roach Coye's current scholarship looks at the intersections of climate change, health equity, and how to mitigate the disproportional impact on the Black community through a decolonial, abolitionist, historical trauma lens. Roach Coye works with marginalized communities to develop interventions to increase equity in the natural environment and increase pro-environmental attitudes. Roach Coye hopes her work will enable Black youth and families to access green spaces and non-human-animal interactions to mitigate race-based stress and trauma symptoms safely and meaningfully, while also building regenerative climate futures for marginalized communities.

Savannah Schaufler is a project assistant for "Air and Environmental Health in the (Post-)COVID-19 World" at the University of Vienna and a PhD candidate at the Doctoral School of Ecology and Evolution. She has published with *Anthropological Review* and University of Exeter Press and presented papers at international conferences in the United Kingdom, Germany, Belgium, and Denmark. During her studies at the University of Vienna, where she graduated with honors in evolutionary anthropology, she participated in several inter- and transdisciplinary projects at the intersection of cultural, human, and biological sciences. In addition, she is finishing her bachelor's degree in cultural and social anthropology. She is also a trained paramedic and has worked as a medical assistant for several years.

Nikiwe Solomon is an early career researcher and lecturer working at the interface of science, technology, politics, and urban river and water management in the Environmental Humanities South Centre and anthropology department at the University of Cape Town, South Africa. Her research focus is on how human and ecological well-being and issues of sustainability are entangled with politics, economics, and technology. Nikiwe served as a research fellow in the Seed Box project "Feminist and Anticolonial Approaches to Environmental Humanities and Justice in the Global South" with research focusing on flows—of currents (water and capital), toxics, and cement. Nikiwe currently serves as Cape Town site principal investigator for the project "Critical Zones Africa: South and East Studies" (CzASE Studies) in the Environmental Humanities South Centre, that, for the next four years, aims to develop research on material flows in the Critical Zone that shapes the everyday, and how tracing lived experience can inform governance for more habitable urban and peri-urban spaces.

María Sol Tiverovsky Scheines has a master's degree in Latin American Studies from Universidad Nacional Autónoma de México and a doctoral degree in philosophy. She has been a postdoctoral fellow since 2021 from the National Research Council of Humanities and Science in Mexico at the Benemérita Universidad Autónoma de Puebla. Her research lines relate to subjectivity and resistance in indigenous communities, racist discourses and literature in Mexico, and Jewish immigration in Argentina.

Jeevan Stephanie Kaur Toor is a PhD student at University College London at the Institute for Global Health where she is researching Inuit youth qanu-inngitsiarutiksait (health and well-being) in Winnipeg. Prior to her doctoral studies she was working with a team from the University of Manitoba on a joint project with community partners, such as the Manitoba Inuit Association, to increase Inuit-specific health care in the region, as currently there are no culturally appropriate services. Her original ancestry is from the Punjab region in India, but her nationality is British Canadian. Jeevan qualified with an MSc in international development from the University of St. Andrews and a BSc in anthropology from the University of Durham. She also contributes to the Arctic Institute's weekly newsletter and is an editor for the Society for the Anthropology for North America's *Anthropology News*.

Index

Note: Page numbers in *italics* refer to illustrative matter.

A

abolition ecologies, 223–24. *See also* policing;
    Stop Cop City movement
action-network theory (ANT), 48–49,
    52–53, 60–62
Adivasi communities, 40, 45n45
air and kinship, 5–6. *See also* air-breath
airborne plastic, 57–60. *See also* air pollution
air-breath, ix, 23–41. *See also* air and kinship;
    lungs
air of freedom, 5
air pollution, ix, 3–19, 57–60, 171, 172
airstories, 13
alcohol use, 84, 85
All That Breathes (Sen), 31
alterlife, 30, 32
Anthropocene, 174

anti-Blackness, 9, 12, 13, 26, 40, 152, 216. *See*
    also racism
Arctic, 73–97
Arctic Council, 92
"Art of the Pandemic" (Global Health), 33
Atlanta, GA, 131–58
Atlanta City Detention Center, 145
Atlanta Community Press Collective, 157
Atlanta Police Foundation, xi, 132, 157
Atlanta Solidarity Fund, 156
atmoactivism, viii, 4, 14–19
atmoracism, 3–19
avatittingnik kamatsiarniq, 79–81

B

Black civic leadership, 144–48
Black Lives Matter movement, 132,

144–48, 150–51, 223. *See also* civil rights movement
breathing, 5–14
Brennan Center for Justice, 150
Brooks, Rayshard, 133, 146

C

Canada Health Act, 94–95
Cape Flats Dune Strandveld, 169
CapeNature Conservation, 169
CDC (Centers for Disease Control and Prevention), 141
Census of Law Enforcement Training Academies (CLETA), 155
civil rights movement, 135, 147, 150–51, 153, 223. *See also* Black Lives Matter movement
Civil War (U.S.), 134, 147
climatic impacts: of COVID-19 pandemic, 23–25, 30–31; on qanuinngitsiarutiksait, 83
clothing waste, ix–x, 54–57
colonialism: in climate crisis, 211–19; health and, 11–12; molecular, ix; and qanuinngitsiarutiksait, 81–83; waste, 50. *See also* decolonizing environmental justice; racial capitalism; slavery
"Colonisation Is a Pyramid Scheme" (Partridge), 81
Comida pa' los Pobres (*Food for the Poor*) (film), 14, 16, 17
Commission for Racial Justice, 185–86, 196
community policing, 131–58
complex personhood, 77–78
Conviction (film), 6
Cop City, xi, 131–32, 138–39, 156–57. *See also* Stop Cop City movement
cosmologies, 220–21
Council on Criminal Justice, 150
COVID-19 pandemic: air pollution and, 170, 172; atmoracism and, 9; effects on nature of, 23–25, 30–31; funeral pyres during, 38–40; grassroots strategies during, 114, 126–27; medical oxygen shortage during, 34–37; political aesthetic of, 26–27; as social triage, 42n11; socioenvironmental sustainability and, 107–27; South African land invasions during, 169–71
crematorium, 24–25. *See also* funeral pyres
Crowd Counting Consortium (CCC), 150–51

D

Dalit communities, 39, 40, 45nn45–46
decolonizing environmental justice, 207–25, 231–33. *See also* colonialism
DHT Holding, 171
diseases, 9, 87–88. *See also* COVID-19
dispossession: abolishing, 225; Black, and gentrification, 152; colonialism and, 207, 211, 213; and COVID-19, 27, 40; and environmental injustices, 207, 222; in South Africa, 170–71. *See also* forced relocation; land invasions
Driftsands Nature Reserve, 169–70, 174
drug use, 84, 85

E

Environmental Concern Scale, 193–94, *195*
environmental connectedness, 193–94
environmental justice, xiv–xvi; decolonizing, 207–25, 231–33; history of, 194–201;

Indigeneity and, 221–22; notions of, 216–17; white-centered, 217–18. *See also* environmental racism

environmental protests, 131–32. *See also* Stop Cop City movement

environmental racism, 7–8, 131–58, 185–89. *See also* environmental justice

environmental sustainability, 115–22

environmental values (EV), 194, 198

erasure, 215–16

ethnoterritories, 115

F

fast fashion, ix–x, 47–62

flower ritual, 124–25

Floyd, George, 12, 15, 133, 144, 146

food insecurity, 85–87. *See also* water insecurity

forced relocation, 73, 82, 88. *See also* dispossession

Freedom Farms Cooperative, 195–96

fueling toxicity, as term, 48

Fulton County jail, 145

funeral pyres, 38–40. *See also* crematorium

future orientation, 199–200

G

George Floyd Justice in Policing Act (2020), 149

Ghost Dance, 112

Great Recession (2008), 152

H

health care access, 141–42

healthy bodies and the environment, 76–77. *See also* diseases

"Healthy Environment and a Healthy Economy, A" (report), 93

heart disease, 9. *See also* diseases

heat island effect, 140

Helicobacter pylori, 87

Hindsight (PBS), viii, 14, 17

I

ice-related accidents, 89–90

India Gate, New Delhi, India, 23, *24*, 25

Indian Act (1876, Canada), 94

Indigenous cosmologies, 220–21

Indigenous Institute of Health and Healing, 75

intersectionality, 186

Inuit Circumpolar Council, 91

Inuit Elders Executive Council, 75

Inuit Qanuinngitsiarutiksait, x, 73–97

Inuit Tapiriit Kanatami (ITK), 94–95

J

jagüey, xi, 107, 109, 122–26

justice. *See* environmental justice

K

Kantamanto Market, ix, 47, 50, 54

King, Martin Luther, Jr., 147, 151, 152–53

kinship and air, 5–6. *See also* air-breath

Kiowa Sun Dance, 112

Kuils River, South Africa, 169–74, *175*

L

land invasions, 170–71. *See also* dispossession

LaPlace, Louisiana, 10

Lown Hospitals Index for Social Responsibility, 142

lung disease, 9. *See also* diseases
lungs, 4, 18, 33, 36, 37, 39, 43n27. *See also* air-breath

M

material flows, 169–73
media, as concept, 43n22
medical oxygen, 34–37, 44n31, 44n35
medicinal plants, 115–22
mental health, 74, 84–85
metabolic rift, 177
Missing Magic (film), 14, 16
Modi, Narendra, 24, 35
molecular colonialism, ix
"Moved" (Rasmussen), 83

N

National Council of Evaluation for Social Development Policy (CONEVAL), 113
necropolitics, xii, 131–58. *See also* COVID-19 pandemic
Ngigua Historical Theater Community Company, 123–24
NoDAPL movement, 132
noise pollution, 171
Now Let Us Sing (film), 14–17

O

Occupy Wall Street, 132
One Human Family Choir, 14–15
oxygen. *See* medical oxygen

P

Páez Terán, Manuel "Tortuguita," xi, 131
Partridge, Taqralik, 76
Pawnee, 112

PEB (pro-environmental behavior), xiii, 184, 198
personhood, 77–78
Plains Indians, 112
PM CARES Fund, 35
poetry, 76
police brutality, xi, 4, 9, 13, 131
policing, xi–xii, 131–58. *See also* abolition ecologies
POPs (persistent organic pollutants), 90–92
protection motivation theory, 198

Q

qanuinngitsiarutiksait, x, 73–97

R

racial capitalism, xiii, 26. *See also* colonialism
racism: atmoracism, 3–19; environmental, 131–58, 212–15; risk and, 185–89, 196–98. *See also* anti-Blackness; colonialism
rain rituals, xi, 107, 109, 122–26
Rasmussen, Knud, 76, 83
remote work, 137–38
residential schools, 73, 88
residential segregation, 8
respiratory problems, 3–19
right to comfort, 134–39
risk and race, 185–89, 196–98

S

Sacrifice Zone, The (film), 6
Sacrifice Zones (book by Lerner), 173
Safe Streets for All program, 154–55
secondhand clothing market, 51
Seemapuri crematorium, 24–25
Selma-to-Montgomery march (1965), 151

Sen, Shaunak, 31
settler colonialism. *See* colonialism
Siddiqui, Danish, 24
Sky media, 28
slavery, 5, 11, 26, 134. *See also* colonialism; racial capitalism
slow violence, ix, 13, 55, 173–77
SmartICE, 90, 97
Smithsonian Science Education Center (SSEC), 183, 189–90
Smithsonian Science for Global Goals, 183, 190, 192, 200
Smoke Signals (film), 40
socioenvironmental sustainability, 107–27
soil contamination, 171–72
space, governance of, 177–78
Standing Rock, 132
Stop Cop City movement, 133, 144–48, 157–58. *See also* abolition ecologies; Cop City
suicide, 74, 84
Sun Dance ceremony, 112
sustainability, 218–19

T

Teatro histórico Ngigua, Xra Juinche'e xan, 123
textile pollution, ix–x, 47, 54–57. *See also* fast fashion
They're Trying to Kill Us (film), 9
This Body (film), 14, 17
toxic exposure, 169–73
toxic fumes, 57–60
toxic timescapes, 55–56, 57
Toxoplasma gondii, 87
traditional ecological knowledge, 112
Tuskegee Institute experiments, 198

U

Udaan (Soar) (film), 14, 17
Unbreathable: The Fight for Healthy Air (film), 6
United Nations, 88, 91
urban-rural divide, 139–44
urban spaces, 139–44
U.S. Department of Homeland Security, 154
U.S. Environmental Protection Agency (EPA), 140

W

waste colonialism, 50. *See also* colonialism
Wasteocene, 52
waste regimes, ix, 51, 52
waste relationality, 51–54
water insecurity, 88–89, 122–26. *See also* food insecurity
Wellstar Health System, 142
We Stay in the House (film), 14, 16
"white climate crisis," xiii–xiv
white eco-justice, 217–18. *See also* environmental justice
white flight, 134–39
WHO (World Health Organization), 58, 79
Why Is Covid Killing People of Color? (film), 9
workplace conditions, 137–38
World Bank, 58

Z

zones of sacrifice, 173–77